THE CARBON CRISIS

Presenting

The real cause of the financial crisis

The three carbon conspiracies

The Green House warming myth

The Carbon Dioxide paradox

Why the Electrical Car is a Joke

*The only real problems we have are corrupt
scientists and incompetent government.*

By John Lincoln

*Author's son & girlfriend Yusi at the edge of the rim of fire in Java, Indonesia
where Carbon Dioxide is returned to the atmosphere after existing as sea bed
limestone for millions of years.*

DORRANCE
PUBLISHING CO
EST. 1920
PITTSBURGH, PENNSYLVANIA 15238

The contents of this work, including, but not limited to, the accuracy of events, people, and places depicted; opinions expressed; permission to use previously published materials included; and any advice given or actions advocated are solely the responsibility of the author, who assumes all liability for said work and indemnifies the publisher against any claims stemming from publication of the work.

Dorrance Publishing Co
585 Alpha Drive
Pittsburgh, PA 15238
Visit our website at *www.dorrancebookstore.com*

ISBN: 978-1-6393-7051-1
eISBN: 978-1-6393-7840-1

About The Author

John Lincoln is an oil and gas entrepreneur and founding father for five innovative international engineering procurement and construction firms. His companies are world leaders in Subsea pipeline and cable technology, natural gas processing, and alternate fuel systems for automobiles. John is also an energy and Earth scientist.

John was born in Miami, Florida in 1950 and graduated from Florida Atlantic University as an ocean engineer. He moved to Australia in 1980. In 1992, he expanded his businesses globally and currently resides in SE Asia and travels to thirty countries a year. John has always had an in-depth intellectual curiosity about the world and is able to understand with great clarity the complex intricacies of any economic, technological, or social systems and also understands how things can be improved. John is well versed in energy systems, climate change, geology, nuclear physics, cosmology, economics, history, spiritualism, and environmental science.

Prologue

This book presents the discoveries, defines and names new trends, and exposes the lies taking place in the world associated with the exploitation of hydrocarbons and its effect on the environment and global economy taking place in the world. These trends include:

The Goldilocks Effect
The Three Carbon Conspiracies
The Green House Warming Myth
The Lincoln Carbon Dioxide Paradox
The Kyoto Protocol Blunder
The First and Second OPEC Wake-up Calls
The Fourth Wave Advancement of Mankind: "The Green Efficient and Renewable Energy Industrialised Revolution."
Outrageous lies being promoted about the Electric Car

Presented in this book is the science of the chemical cycle of carbon and the roll of the ocean and limestone in removing it from the atmosphere. The amount of carbon on earth is presented for the first time and the percentages in the ocean, limestone, plant and animal matter, hydrocarbons and earth's atmosphere are established. The history of carbon utilisation from the use of coal during the Industrial Revolution to the discovery and utilisation of oil and the growth of the major oil companies, the economic havoc wrecked by OPEC in the seventies to the economic meltdown of 2008.

Green-house warming supposedly caused by CO_2 is firmly proved false. In fact, it is proved the converse is true, that is the warming earth is producing the CO_2. Much like CO_2 bubbling out of a carbonated beverage left out on the counter.

The future of mankind for the next 200 years is predicted and his adaptation to a new economic system centered around; reduced energy consumption, reduced exploitation of natural resources, reduced populations, migration to renewable energy centers and global "One World" integration of cultures and religions.

The lies about the Electric car saying it is running on clean energy is about corruption as Biden, Pelosi and Schumer are hoping for millions of dollars of payments under the table by Elon Musk of Tesla. The fact is electricity is mainly from the dirtiest of all hydro carbons coal with a carbon footprint 50% greater than natural gas. Electricity as of 2019 is also 40% more expensive than gasoline and 300% more expensive than natural gas or methane. There is over 300 years supply of methane and all cars can be converted to run on methane or gasoline for about $1500. This fuel should be receiving the government's endorsement

John Lincoln returned to the USA in 2014 and has lived there until now 2021 when this book was updated. It has been twelve years America has recovered a little due to cheap oil but is still poorer than it was in the 2000s. The carbon content in the atmosphere has increased from 380 ppm to 400 ppm or about 5% in 28 years not much and if it doubled say 100% it would only give us a greener planet.

During the Obama administration there were more handouts for the poor, and our businesses continued to fail as unfavourable tax structures caused companies to manufacture in China and Mexico. Our balance of payments with China was excessively negative as we imported trillions of dollars of their goods. Our Veterans were mistreated. Donald Trump was elected and things started to improve: we taxed China, took care of our Vets, closed our boarders to criminals, pulled out of the Paris Accords and quit wasting money on alternative and fake clean energy. Now we have Joe Biden uneducated in energy and the dirty energy feeding the electric car trying to get us all on alternate energy or dirty coal as it powers the electric car for which he may receive millions of dollars of bribes.

Contents

1.0 INTRODUCTION

Something happened between 20,000 BC and 10,000 BC when mankind walked from the caves as the earth warmed and the glaciers retreated and the cold white ball called earth started to become green and proliferate with plants and animals. In 1987–2008 at the peak of the proliferation mankind started to worry that the earth was getting warm and there was an increase of carbon dioxide (CO_2) in the atmosphere. Of course, it has warmed somewhat since then and there is more, although a small amount, of CO_2. If it wasn't warm and the increase in CO_2 we wouldn't be here to notice. This is called the "Goldilocks Effect" where everything is just right. We need warmth and CO_2 for our plants to grow so we can eat.

It is one of the intentions of this book to substantiate that all the greenhouse warming hype proposed to be caused by CO_2 is nothing more than false propaganda being spread by the politicians of western countries that 1) have no oil to stop China and India from using all the excess oil which is creating havoc in the western economies 2) profit from an electric car industry. This is the "third carbon conspiracy."

It is also the intention of this book to present the first quantitative analysis to determine where the CO_2 and potential CO_2 from the oxidation of carbon actually are and at what quantities. As will be proved over 9.25% is dissolved in the ocean, about 90.7% is locked in the world's limestone and the rest about .05% found in living plant and animal matter, underground as oil, coal and gas and in our atmosphere. Our atmosphere has as a mere 7.13 parts per million or 0.000713 per cent (almost nothing) of the total CO_2 and potential CO_2 should the earths plants, animals and hydrocarbons be oxidised on Earth.

This book also presents the first quantitative analysis to show how much CO_2 is actually placed in our atmosphere each year from the burning of hydrocarbons which is about 32.5 trillion kilograms and how much the earth's plants especially the oceans plants as mostly phytoplankton remove each year about 300 trillion kilograms or ten times what mankind produces. Why then has our CO_2 content in the atmosphere increased 100ppm in 140 years or about 0.187% per year? A major discovery made in this book is that it is most probably either coming out of the ocean just like the CO_2 bubbling out of a served carbonated beverage warming to room temperature without its top on or increased volcanic activity discharge of CO_2 or a combination of both.

The western powers, the largest consumers of oil just happen to be the countries with no or little oil. These countries include the United States, England, France, Germany and Australia. Many of these countries have been typically unfriendly to each other and only recently in (2005) started cooperating together (the first carbon conspiracy) in the Middle East to destabilise Iraq, Iran and assist in their common interest which is to survive the recent oil shock price increases, which is destroying their economies. The oil price increase caused by OPEC mainly in the Middle East through its affiliate in the Middle East the Gulf Cooperation Council (GCC) is the "second carbon conspiracy." The USA consumes 20 million barrels per day of oil of which 50% is imported. The total amount of US oil in storage is 300 million barrels plus 700 million barrels in the Federal Governments Strategic Petroleum Reserve. This equates to fifty days' supply. An interesting fact is in 1980 when I left the USA there were only 200 million people (now there is about 300 million) and the oil consumption was the same at 20 million barrels per day the importation rate was about 80%. This means the USA has come a long way in finding more oil mainly in the deep waters of the Gulf of Mexico and also in reducing consumption by switching to coal for power and also better practises at conserving energy like better home and building insulation.

The average American family of four in 2008 paying US $7,000 per year to import oil. This combined with the war in Iraq, a similar debt, will put the average American family in debt by US $161,000 after the eight-year presidential term of George Bush. Quite a legacy for an American president. When this book was first written in 2008, the housing and stock markets have crashed, 30%–40% and the price of homes have crashed 50% in some cities, several banks, lending institutions and mortgage insurance companies have declared bankruptcy as a result. How could leaders responsible for the stability of their country irresponsibly let this happen? In 2004, when asked by a news reporter, did it "matter to Americans' wealth that they were not saving?" Alan Greenspan replied, "The price of their homes was so high they had real wealth in that and so saving was not important." Luckily, he resigned a couple of years later, wrote a book and tended to distance himself from the financial catastrophe he helped create. So, with an economic recession now in place with interest rates rising, what does Bernanke the new leader of the Federal Reserve do, drop interest rates the same thing that helped get everyone in the mess to start with.

The effect will not work as home prices have fallen 50% in most major cities; many homeowners actually owe the banks now more than what the home is worth. The result, the owners simply walk away or default on the loan and let the banks have the property. The bank sells for a loss. The banks cannot afford to follow the Federal Reserve's lead and lower interest rates.

Citibank facing financial disaster has sold most of its shares to China, Saudi Arabia and the Singapore government. Bear Sterns nearing bankruptcy as well as the lending institutions Freddie Mac and Fannie May and mortgage insurer American International Group (AIG) have been rescued by the USA Federal government. The public will then have to foot the bill by paying taxes for the irresponsibility of the greedy bank and loan officers and incompetent government regulators that are supposed to keep this from happening.

The American stocks fell in early 2008 because of investors panic selling their shares. Selling is based on a belief that they are over inflated and will fall anyway. Europe, Australia, Singapore, India, Chinese and Malaysian share markets have also crashed and followed the Americans.

By March 2008, the American dollar, as a result of the Federal Reserve dropping the interest rates has dropped through the floor at an all-time low down about 50% of the Australian dollar. This means Americans and even international companies with contracts to be paid in US dollars will have to pay much more for foreign services or importing items and also will get paid less. Gold is also going through the roof almost US $1,000 per ounce and the OPEC countries are also raising the price of oil to compensate for the weak US dollars they receive for it.

The problem now is even though there is more oil out there, the new demand for oil from China and India far exceeds what can currently be produced. Don't rely on the oil companies this time like they did in the late 1970s to produce more oil and stabilise the world's economy and prevent a major recession hanging over head like the great depression.

The level of carbon dioxide in the atmosphere is .038% which is 380 parts per million. All of us have had 380 dollars at one time or another but hardly anyone has had a million dollars. If you drew a large circle, the finest line from the centre out would represent the current level of CO_2 in the atmosphere. The level CO_2 is just too small to worry about and has occurred at similar but slightly less level at times over the last 400,000 years as preserved in the Greenland ice cores.

CO_2 even at much higher concentrations has almost no effect anyway, less than 5% in trapping the earth's radiant energy sent back into space. Hyperbaric medical technicians and I worked as a deep diver, supporting the lives of deep divers and astronauts are very much aware that the body can tolerate much higher concentrations of CO_2. In fact, over 20 times greater and at this rate our deserts would turn into jungles and forests similar to when the dinosaurs roamed the earth. The reason the dinosaurs were so big is that there was lots of plants and animals to eat. The ocean removes most of the CO_2 and turns it into limestone and coral on the seabed, which it has done for hundreds of millions of years. This is why there is a limestone layer over 1200 meters thick over most of the earth's warm latitudes.

From the Greenland and Antarctica ice cores it is apparent the earth's temperature is constantly changing. In fact, the earth has been mainly a ball of ice with only four warming periods in 400,000 years, and with these warm periods then was high CO_2 concentration. It does not take Albert Einstein to work out mankind did not burn hydro carbon back then and so Global Warming associated with CO_2 can only mean it's the global warming that causes the high CO_2. Think about it if the earth is a ball of ice, what animals are there to produce CO_2 or forests to burn. None. This is the "Lincoln carbon dioxide paradox." It is the earth's warming that produces the CO_2 not the other way around.

So how come some people are promoting the 'Greenhouse Gas Warming Myth' and blaming it all on CO_2? Well, we can see a very slight warming trend that has increased since 1980, we are told the glaciers and the polar ice caps are melting and we can see millions of cars and thousands of power plants pouring CO_2 in the atmosphere. Why not blame it on the CO_2 it seems like a good excuse. Governments, companies and especially the media can easily control you and sell you products and services by creating fear and despair. This is the "third carbon conspiracy." Responsible scientists that look for the truth know that global warming is not only normal but also essential for us to live our modern lifestyles.

Attempts to reduce our dependence of non-renewable energy such as oil, gas and coal will prove beneficial but will require major revolutions in the living habits of mankind, probably not achievable in the foreseeable future to be completely independent of hydrocarbons. But do not worry we have hundreds of years left of natural gas and coal. Natural gas or methane is a natural occurring substance in the universe and also can be produced from the anaerobic decomposing of organic matter. Many garbage dumps worldwide generate electricity from taping the methane given off. At home a family of four can actually make enough biogas from the anaerobic decomposition of table scraps and human solid waste to drive a small car up to 20 kilometres per day. Today (2020) in America Electrical Power is produces 40% from coal, 27% from methane gas, 27% nuclear and 6% from hydroelectric and wind. Discussions in 2020 centre around electrical cars however electricity cost twice what gasoline cost and if everybody was to have electric cars the cost of upgraded power plants, transmission lines and facilities at your home would cost many trillions of dollars probably raising electricity costs by a factor of three or about twelve dollars per equivalent gallon of gasoline.

2.0 THE CARBON DIOXIDE CYCLE

2.1 Oxidation

Carbon is the most basic element for life, in pure form it is a black brittle solid at normal temperature and pressure condition. Carbon is the most important element on earth as when it combines with oxygen it releases huge amounts of energy and also carbon dioxide. All animal and plant matter dead or alive contain carbon. Carbon dioxide or CO_2 is the clean perfect combustion or oxidation bi-product of fossil fuels in engines and furnaces and also as a result of the combustion of organic matter like trees and animal flesh. In a similar process although on a slower less energy intensive reaction is the oxidation (combining with oxygen) of organic matter by animals and some bacteria called respiration. Every time you breathe you take in oxygen, the chemical process of respiration, a form of oxidation takes place in your billions of cells and carbon dioxide, and energy is released.

2.2 Photosynthesis (Carbon Dioxide is removed)

The opposite reaction to oxidation, which is thousands of times greater is photosynthesis and all the earths plants and especially the oceans phytoplankton's which comprise about 80% of earths plants absorb energy from the sun, remove the carbon dioxide and produce oxygen and organic carbon-based plant material which can be eaten by the marine animals, bacteria and zoo plankton.

About 150 watts per square metre of direct sunlight on a average day reaches earth. If you calculate half the earth's surface area based on a diameter of 12,756 kilometres and that you are receiving the energy 24 hours per day, you will come up with 1.44×10^{23} Calories (one hundred and forty billion trillion calories hitting the earth each year) of this about 40% is reflected from the clouds, ice and ocean reflection. 53.39% is used in the evaporation of the oceans, which provides our rain, snow, cooling of the oceans, and generation of low-pressure storms that assist with the global wind and ocean circulatory currents, 6.3% is used to heat the air and help create the global winds. The balance, which is 0.31%, the most important part, is utilised as biological fixation from the plants on earth mainly in the oceans during photosynthesis giving us oxygen and food. About 4.4×10^{20} calories per year or 440 million trillion calories are fixed this way. Based on 88 trillion kilograms of plants produced in the sea each year (Wiley Ocean Engineering University of California Text Book 1968) allowing an additional 20% for land plants this equates to about 11×10^{13} or one hundred and ten trillion kilograms per year of plant and animal mass-produced. The amount of CO_2 removed from the atmosphere per year based on 2.75 kilograms of CO_2 per kilogram of animal and plant matter is 3.0×10^{14} or 300 trillion kilograms of CO_2.

2.3 In the Ocean 9.25% of Earths Carbon

The surface area of the oceans comprises 75% of the earth's surface. The average depth is 4000 meters with a hydrostatic pressure 400 times that of our atmosphere. The temperature drops rapidly from the surface and gradually cools to about 0°C. If all the land masses were rounded off the mountains levelled and the ocean trenches filled then the earth would be nothing more than an entire ocean about 3000 meters deep full of cold salty water loaded with CO_2. The cold water and the great pressure at depth holds the CO_2 in great quantities like a bottle of a carbonated beverage with the cap on in the refrigerator. Only thing is the immense pressure at the bottom of the ocean is over 400 times greater than any carbonated beverage thus providing enormous CO_2 storage.

Oceanographic scientists report a dissolved CO_2 content in the form of carbonic acid of comprised mainly of bicarbonate HCO_3^{-1} and carbonate CO_3^{-2} 28mg per litre at the ocean surface. Because of the increased compressibility and solubility of a gas under immense ocean pressure if we assume the initial CO_2 gas which causes the carbonic acid behaves like an ideal gas, this increases on average to about 10.2 grams per litre at a depth of 4000 meters. Taking 75% of the surface area of the earth times 4000 metres by this concentration of CO_2 yields a total weight of 17 x10^{18} kg or 17 million trillion Kg of dissolved CO_2 in the oceans.

Carbon dioxide is not however an ideal gas like nitrogen, oxygen and methane. These gases cannot be compressed into a liquid without substantial cooling to temperatures down to -170°C. Carbon dioxide or dry ice as we also know it can easily be turned into a liquid or solid from the addition of pressure alone. At a depth of 730 meters or 1,197 psi carbon dioxide becomes a liquid.

Phase diagrams for pure substances like water, CO_2, methane, oxygen, nitrogen, propane, butane and hundreds of other compounds can be produced in a laboratory. Here the various temperature and pressure conditions can be established for the physical phases of the substance. Water for example at atmospheric pressure starts to boil and convert to steam, a gas at 100°C and until all the water boils into steam the temperature remains constant. At a lesser-pressure say like a vacuum it will boil at normal temperatures. A phase diagram for CO_2 can easily be obtained from any engineer, chemist or the internet. From inspection of a phase diagram of CO_2 here you can see for an average ocean temperature profile at the depth of 340m has a temperature of 8°C it can start to turn into a solid (dry ice). In sea water however it forms hydrate which is like dry ice but has sea water and some methane sometimes dissolved in it.

From inspection of a phase diagram, you can see that at a depth of 2500m and a temperature of 2°C CO_2 becomes the same density as seawater namely 1.04 kg per litre. If colder or deeper it will sink deep in the ocean. Therefore, CO_2 can exist in much higher concentrations in hydrate form in the deeper colder waters. This hydrate can exist on the bottom of the ocean like water sitting in lakes on land. It can also be dissolved in the deep ocean waters just like water is dissolved in our atmosphere. And similar to the water in our atmosphere if the temperature or pressure falls you can get retrograde condensation where like water falling out as rain or snow in this case liquid or solid CO_2.

Unlike our atmosphere which is a gas and so when it gets hotter it dissolves more, water behaves the opposite at warmer temperatures it will hold less dissolved substances. An exception to this is calcium carbonate or limestone which dissolves more at colder temperatures.

So, where there is significant upwelling like off the west coast of South America, one of the best fishing grounds, in the world, cold-, nutrient-, and CO_2-rich waters surface to feed the phytoplankton and stimulate large fish growth up the food chain. CO_2 could also easily come out of solution just like rain except it bubbles upward when encountering the warm shallow water and helps contribute to the atmosphere levels. On the other hand, warm CO_2 rich waters are carried by many ocean currents like the Gulf Stream into the colder Northern waters, which sink off Ireland and England and return the CO_2 to the deep depths.

The amount of CO_2 trapped in the colder deeper ocean waters has never been investigated very much. Being out on the deep ocean and having equipment capable of going down, 4000 metres is very expensive. As an ocean engineer and my company running several deep-water oil and gas operations, I know the costs are in the order of fifty to 100,000 US dollars per day. Doing tests in situ would be best. Recovering water samples in heavy steel high-pressure cylinders is possible but difficult to test once on the surface and in addition the chemical reaction of the contents with the steel cylinder can alter the findings.

How much CO_2 is trapped in the dissolved hydrates or sitting on the ocean floor no one knows. Experimental researchers have dissolved liquid CO_2 in high pressure and cold salt water (simulated sea water) at concentrations of up to 50grm per litre.

This is five times greater than the concentration would be based on the ideal gas calculations just done however this included all of the ocean water Colum. If we take 4000 metres as the average ocean depth and know hydrates or liquid CO_2 cannot exist above 2500 metres, we have a water Colum 1500 metres deep which could have highly concentrated CO_2. As an approximation, which is really no more than an educated guess, I am going to assume the concentration is 25 grams per litre or an additional 14.2 grams per litre for 35% of the water column. This equates to increasing the CO_2 in the ocean from the theoretical 17×10^{18} kg by 50% to 25.5×10^{18} or 25.5 million trillion kilograms of CO_2 in the oceans.

2.4 Seabed Limestone 90.7% of Earth's Carbon

Zoo plankton, which are tiny almost microscopic animals in the sea like Oolite and Foraminifera comprise about 80% of all Earth's animals. These animals consume the phytoplankton, exhale carbon dioxide and the rest of the carbon is left in their skeletal frames in the form of calcium carbonate or calcite. Luckily there are much more floating plants in the ocean than animals and as a result the land-dwelling animals like us have fresh air to breathe. There is a net production of oxygen and hence a net absorption of carbon dioxide into the sea which ends up in the animal skeletons. These sea animals' skeletons pile up on the sea floor particularly in the warmer latitudes and form limestone which is the basic building material bed rock. Corals and animals when they die also leave a calcareous skeleton. This limestone and calcareous sands, shells and corals comprise the bed rock of most tropical and subtropical above and below water land masses. The whole state of Florida, the Bahamas much of Mexico, all the Caribbean islands, North and South America and all the tropics of Africa, Asia, Australia and Europe have limestone bedrock. This limestone layer is up to 4000-feet deep at places. Many temperate regions of earth also have limestone bedrock or marble like in the USA and Europe indicative of continental drift. Since the limestone and marble can be found as far as 1200 kilometres north of temperate regions and if we assume an average continental drift of 75mm per year we know this process of removing CO_2 and converting to limestone has been going on for at least 16 million years.

Geologists that study the earth's rocks know photosynthesis has been occurring for more than a billion years. The Grand Canyon, as can be seen by an internet search for "Grand Age of Rocks Grand Canyon," is perhaps the best deepest record of the deposit of limestone and other sedimentary rocks and is still there for anyone to go see. The Columbia River has cut down and exposed 3,825 feet of mainly ocean deposited rock cliffs all the way down to base igneous rock 3.5 billion years old. The age of rocks can be established by radioactive dating techniques especially using accurate carbon dating methods for organic deposits like shells and other fossils left on the sea bed back then. The types of animal fossils found gives excellent evidence to what type of animals lived then and is a good basis for geology and paleontology. Alternate bands of limestone with other sediments like sand, and slate indicating a dry time above water also provide excellent records of when the area was covered with ocean and the ancient levels of the ocean.

An internet search for "Florida Aquifer System" figure 52 as a result of extensive drill coring indicates most of Florida has limestone down to 4,000 feet deep the oldest deposited during the Paleocene Era 66 million years ago. Other searches around the world can be done but drilling is expensive and investigating sea ridges is also expensive and would probably not indicate any exposed limestone deeper than 2000 meters since here it would dissolve because of the cold temperatures as discussed below. Does that mean in Florida, our East Coast is being undermined by cold water down 6560 feet (2000 meters)? No, as at that depth only igneous or other non-calcareous rock can be found. Besides the Florida Straits is only 396 meters deep so we are safe.

Calcium carbonate ($CaCO_3$) or limestone is another substance on earth which has an anomaly unlike any other substance in that its solubility in water increases with a decrease in temperature. Because of this it does not exist in solid form in the deep colder depths of the ocean and can only be found on land and down to a maximum ocean depth of 2000 metres.

In warm water it can easily be assimilated in the skeletons of marine microscopic animals and is found as bed rock extensively on the tropic and sub tropic regions of the earth from +30° latitude to -30° latitude. It is also found as metamorphosed marble in Europe outside of these latitudes probably formed millions of years ago in the tropical regions since then continental drift has moved it there or the area was submerged at one time.

If you take about 65% of the surface of the earth the area between -30° and +30° latitude times the area of land mass and submerged areas less than 2000m water depth you end up with 33% of the earth's surface.

Multiply this by an assumed average depth of 1200 metres yields about 2.0 x 10^{17} cubic metres of limestone or 200 thousand trillion cubic meters.

The mass of limestone is 2400 kg/m³ and so the total mass of limestone in this area is 4.8 x 10^{20} kg. Taking the mass ratio of CO_2 in $CaCO_3$ as 44% yields 2.1 x 10^{20} kg of CO_2. Increasing this say 20% to allow for other areas of the earth where limestone was previously formed when its land mass was in the warm regions yields 2.5 x 10^{20} or 250 million trillion kilograms of CO_2 stored in limestone on earth. Interesting to note if one assumes say the average depth of limestone is say only 800 meters or 33% less thick if you re do all the calculations the percentage of the total CO_2 is 87% of the earth's total not 90.74% a mere 4% difference. Of course, if the average depth is way beyond 1200 m say 1600 m thick the percentage of the total CO_2 is 93%. The point is no matter what reasonable thickness you assume the percentage of CO_2 in the limestone sediments is very large... say 87 to 93 percent of the world's total.

2.5 Hydrocarbons Underground (0.0023% of Earth's Carbon)

Organic matter like leaves and tree trunks get broken up and eaten by insects, termites, and bacteria. The resulting detritus is washed into the streams and rivers by the rain and out to sea where it piles up with sediment and forms the great ocean deltas which are up to 10,000-feet thick. Under pressure, this organic matter turns into oil, gas and coal. It has been estimated or roughly guessed by Wiley in the Ocean Engineering University of California Text Book 1968 there was approximately 2.2 10^{22} calories of hydrocarbons left underground. Wiley also estimates an absolute maximum amount of hydrocarbons energy we could possibly exploit would be 4.4 x 10^{24} calories or 200 times what is believed to exist, as this figure would deplete our planet of its current oxygen content.

If we work on 2.2 x 10^{22} calories of hydrocarbon energy left and allow about 44 M.J (mega joules) of energy consumed per kilogram of CO_2 produced, we get 2.1×10^{15} kg or 2100 trillion kg of CO_2 that could be put in the atmosphere by the burning of all the existing known or predicted hydrocarbons.

Approximately 87 million barrels of oil are consumed each day in the world. Allowing the barrel is 159 litres and oil weighs approximately .85kg per litre yields a mass of 4.2×10^{12} kg of oil used per year. Oil like most hydrocarbons except natural gas or methane which produces less CO_2 produces about 3 times as much weight in CO_2 production as the fuel that was burned. This equates to about 12.5×10^{12} kg of CO_2 per year is produced from the burning of oil. If we allow the fact oil makes up 43% of all the hydrocarbons burned, coal 31% and gas 26% and assume the same ratio of 3 being the weight of CO_2 produced to hydrocarbons weight burnt we get a total discharge into the atmosphere of 3.25×10^{13} or 32.5 trillion kilograms of CO_2 per year.

Interestingly enough, if we divide this discharge rate into the total amount of carbon fossil fuel underground, we get about sixty-five years of consuming hydrocarbons at the same rate as in 2007. Since the Wiley estimate of total hydrocarbons was made in 1968, we can estimate about twenty-five years of hydrocarbons fuel left. Interestingly, also is that if there was more in the ground, much more, we could burn it for 30,800 years before we used up all our oxygen. Fortunately, there appears to be much more in the ground than predicted by Wiley more discoveries are made each year and more efficient extraction methods like fracking a method involving pumping clean high-pressure water into shale causing the release of significantly more methane. We will, however, use Wiley's number times 3, allowing for 6300 trillion kilograms of hydrocarbons left underground.

2.6 Earths Plants and Animals 0.0016% of Earths Carbon

According to Wiley Ocean Engineering, about 8.8×10^{13} kilograms of plants are produced in the sea each year from photosynthesis hence the removal of carbon dioxide and production of oxygen.

If we increase this 25% to allow for the land's plants, we get about 11 x 10^{13} kilograms of total dry plant (no water included) mass-produced. Some of the plants are eaten by animals but this is the total gross production of plants or net production of plants plus net increase of the mass of animals which should be the same.

If we allow taking the ratios of the molecular weight of CO_2 which equals 44 and an approximate weight of the simplest hydrocarbon CH_2O, which is 16 we get 2.25 times more CO_2 mass removed than the original mass of the plants. We then calculate 30×10^{13} or 300 trillion kilograms per year of CO_2 removed. Interestingly to note the earth's plants remove about ten times the amount of CO_2 of what man puts into the atmosphere?

An assumption made now which may be an approximate guess is that the production of plants and animals each year is six percent of what exists. On this assumption the total weight of plant and animal matter on the planet is 18.3×10^{14} kilograms which is then sixteen times what is produced, of course assuming the total mass remains the same the same amount of animal and plant produced each year also is eaten and dies and if it has a carbonate skeleton is recycled back to the sea or limestone. Allowing of the same ratio of 2.5 being the amount of CO_2 that could be entered into the earth if all the plants and animals were oxidised would be 41.2×10^{14} or 4120 trillion kilograms of CO_2.

2.7 Earth's Atmosphere 0.000713% of Earth's Carbon

Taking the diameter of the earth at 12,756 km you can calculate a surface area of 5.1 x 108 Km².

On this surface, the pressure is exerted of 1 bar or approximately 14.7 lbs/in² or 10,335 kg/m3. From the weight of all the gases above multiply this area by this pressure and you get the weight of the earth's atmosphere which is 52.7 1017 kg. Allowing for the partial pressure of CO_2 to be reflected of the percentage of its occurrence or 380 ppm or .038%, we get a total mass of CO_2 in the atmosphere of 1967 trillion kilograms. This corresponds to only .000713% of all the earth's existing carbon dioxide or potential carbon dioxide should all hydrocarbons, plant, and animal matter be oxidised.

The CO_2 in the earth's atmosphere has increased 100 ppm in 140 years or an increase of .187% per year or 3.67 trillion kilograms per year. What this means is the CO2 is increasing at a rate of 0.7106 ppm per year and for a maximum level of 20 times existing which is 7,600 ppm this would take 10,695 years assuming everything remained the same. Since we will probably run out of hydrocarbons within 154 years it wont be mankind's contribution to any significant CO2 increase. Mankind also will probably evolve to handle much higher concentrations anyway in over a hundred generations

2.8 CO_2 Makes Planet Green

Well, if you understand what you have just read it doesn't take a genius to show our earth can remove CO_2 much faster than it can be produced. If the CO_2 level falls too much our green beautiful forests turn into deserts. When there is a lot of CO_2 around, our planet thrives because green plants grow huge and there is lots of food. Not really a bad thing. The dinosaurs got big eating all these big plants so the CO_2 level must have been much greater then. Unfortunately, we only have measurable records of CO_2 dating back 400 hundred thousand years in the Greenland Ice cores.

2.9 Green Earth and Red Mars

The discerning reader should now realise it is our carbon dioxide exhausts that is making the planet green maybe too green for our own good. Lots of food, lots of people, and more deforestation and more concrete jungle.

What happens when mankind uses up most of the oil, coal and gas probably within the next one to two hundred years if we continue at the same rate of depletion, the ocean and rivers keep absorbing the carbon dioxide, and so, with no other forces at work, we could in theory turn into a cold ball of dust much like the Planet Mars. Mars it is believed had great oceans we expect to find one day calcium carbonate rocks there. Meteorites have been found from the moon in Antarctica and they show evidence of fossilised life forms. What Mars doesn't have and we do have however an active energetic inner core of the earth is where massive radioactive processes keep the earth's core hot as a molten liquid. This churning magma causes continental drift and the continents and seabed rip apart at places and fold in on each other at other places.

It is this folding in where the global fault lines, seismic activity and volcanoes exist. It is here where the crust of the earth including the oceans bed rock and best of all calcium carbonate rocks get melted and vaporised. The result everything returns to where it came in the molten pits of the earth's core.

2.10 Volcanoes Give Back CO2, Dust and Cool Earth

Fortunately for us, like a giant burp the water and carbon are returned quickly as steam and liquid carbon dioxide from the subsea vents and volcanoes. Sea water salt comprised of sodium chloride and sodium sulphate are returned as chlorine gas, sulphur and sulphur dioxide amongst other things. The rest of the heavy rocks aren't so fortunate unless there is a large volcanic eruption as they normally are very dense and fall deep into the bowls of the molten earth.

Not to get too far off the carbon cycle track but a lot of media phobia about destruction of the ozone layer attributed to under arm sprays releasing hydro flora carbons is actually caused by the release of Volcanic chlorine gas. The largest hole in the ozone layer over Antarctica just happens to be directly over Mount Erebus which discharges thousands of tons of chlorine gas daily. Volcanoes are known to have a much greater effect on the earth's climate. When they erupt, they put billions of tons of pyroclastic debris or rock dust in the atmosphere that causes the sun rays to reflect back into space. Large meteorites have been known to destroy the earth at least twice as they cause volcanos, explosions and also put large volumes of dust in the atmosphere.

The globalisation of the third wave advancement of mankind is now putting strain on our energy and commodity supplies, and a new—as I define it—"Fourth Wave advancement," of mankind, "The energy efficient revolution" is now just starting to take place.

If you listen to and believe (and I don't) these that claim CO_2 is making the earth warm it has taken approximately 100 years of carbon dioxide in the atmosphere to raise the earth's temperature .5 degrees centigrade. One volcano however can lower the earth's temperature ten times this such as Krakatoa, which erupted in Indonesia in 1865. Also, about 40,000 years ago, the earth's temperature was lowered substantially from a single massive eruption in Sumatra Indonesia. Geologists have studied the Foraminifera sediments and found the animals were very small then. In addition, dust and sulphur dioxide sediment can be found in the Greenland ice cores dating back to this time. The dust has been traced back to Sumatra and the once large crater is a giant lake today called Lake Toba.

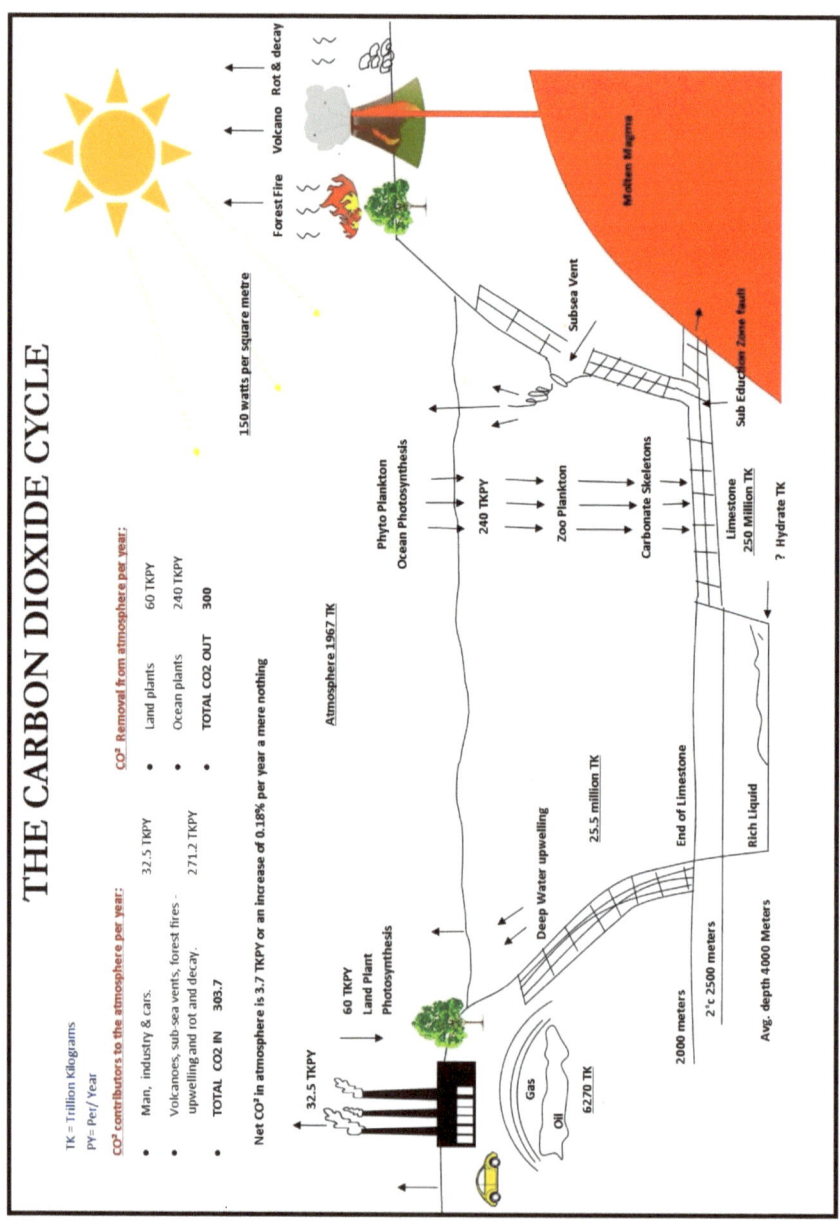

Figure 2.1: The Carbon Dioxide Cycle

3.0 GLOBAL WARMING

3.1 Temperature Recordings Since 1900s

What do we really positively know about the earth's temperature now or even a hundred years ago when it first started to be recorded? Not Much.

If you can believe and rely on the average temperature recording since 1900 from the designated base stations all over the earth it would appear the earth's temperature has increased 0.5 degrees over the last 100 years. The same recordings indicated a general warming of about 0.4 degrees Celsius between 1940 and 1900. After 1940 to about 1980, the earth's temperature cooled 0.4 degrees and since 1980 to 2000 it warmed another 0.5 degrees or about 0.1 degrees hotter than we were in 1940 and 0.5 degrees hotter than in 1900.

If you have studied the effects of volcanoes, you would now become very worried as one volcano, as Krakatoa did in 1886, can put enough debris in the atmosphere to lower the earth's temperature in one week about ten times what it took one hundred years of supposed greenhouse warming to achieve. After Krakatoa, snowfall and frost reported throughout USA and Europe in the summer. This also happened in 1815 and was termed a mini-ice age after the Indonesian volcano Tambora erupted killing hundreds of thousands of people including those that died from the cold and starvation in the Northern Hemisphere.

Another problem is with the methods for recording the temperature. Temperatures at a designated location vary depending upon the time of the year, the time of day, whether it is cloudy or raining and depending on what large air mass moves into the area. Obtaining a reliable average measurement is impossible especially to .5-degree accuracy.

Another problem is the standard scientific method for recording the temperature is to place a thermometer in a standard wood box painted white with louvers to allow the air to circulate. The problem is the box tends to get dark from dust, mildew and not to mention the spider webs covering some of the louvers. The result is the box will always record a warmer temperature as time goes by. The discerning reader will now realise the dust in the atmosphere that reflects the sun's rays and tends to decrease the earth's temperature also gives us false warmer reading on our little white wooden box. This means dust not only makes it colder but our temperature readings are false and should read colder.

Another fundamental flaw with global warming attributed to mainly CO_2 in the atmosphere is why the earth's temperature suddenly dropped between 1940 and 1980. This was the largest ever at the time dumping of CO_2 in the atmosphere from the burning of coal, oil and gas. Not to mention during WWII, where enormous amounts of energy was consumed in the production of steel, manufacture of ships and planes, guns and ammunition and the energy to power and transport these tools of mass destruction. Could it be all the above-ground nuclear testing in the late '40s and '50s and many volcanoes of the times, including Mount Saint Helens in Washington State in 1980—both of which put dust in the atmosphere which reflects the sun's rays—was more influential in cooling the earth. If this is the case then maybe that dust again is settling out and causing the earth to get a little warm again.

It is April 2008; we just had a freak 100 km per hour (60 mph) dust storm from the desert in the areas in the Northwest. Nothing had happened like this in Melbourne Australia before. The dust not only coated the cities, cars, streets and homes but also caused a lot of electrical facility failures. Transformers, generators and electric motors failed through the city as the dust created an electrically conductive atmosphere in many devices causing them to short out. Nobody said a word about the temperature though. It was actually a nice cool wind but from a direction the trees were not accustomed to (from the North) the result was trees, limbs and power lines were down everywhere. Large traffic jams a reminder of worse time to come if we don't get the car off the road soon.

3.2 Greenland Ice Cores

A continuous scientific ice core-drilling programme has been going on in Greenland for forty years and more recently in Antarctica. Each year trapped in the Glaciers is another ice layer deposited each winter.

The ice cores penetrate about 300 meters deep and cover about 400,000 years. In the ice each year, trapped is gases like oxygen, ozone, carbon dioxide, and from volcanoes, yes dust again, that tell-tale nuscience, a science that is proving very important as we can tell when and where dust came from and, combined with the gases, get good forensic evidence of the history of the earth—for the last 400,000 years anyway. A new branch of forensic science of studying dust maybe called Dustology would sure do the world a favour so we can look at all the smoking guns and have better clues before we accuse and convict the wrong culprit.

I am flying over the outback Australia north west of Melbourne as the Boeing 777 flies, looking out the window at eye level about 30,000 feet is bright blue sky above and a red baron sparse vegetation landscape below and guess what is between me and the ground? Dust, 30,000 feet of it. I wonder if it is cooling the desert. I know it must be hot down there and I know everything freezes at night there. It is a cool -42 degrees up here, but luckily, I am in a heated, pressured jet.

From the ice core data, someone—an impartial scientist, I hope—has plotted the estimated earth's temperature and the CO_2 concentration against time for the last 400,000 years. The best easily available representation of this can be found in National Geographic magazine August 2007 issue fold out. Thankyou Australia National Geographic for now apparently presenting a scientifically balanced view on the CO2 and global warming issue. I am not sure if you intended to give the readers the balanced view or not. But from inspection of the graph, it would appear the earth has had three warm periods (warmer than today) over the last 400,000 years and guess what very high carbon dioxide levels with each of these warm periods. It does not take Albert Einstein to realise since mankind was not here releasing CO_2 then and since the only other known way of introducing large amounts of CO_2 in to the atmosphere is from volcanoes which actually lower the earth's temperature it is the warm earth which makes the CO_2 not the CO_2 making the earth warm. This is called the Lincoln carbon dioxide paradox.

Thank you, *National Geographic*. You have now unwittingly proved the CO_2 has little or no effect on the earth's temperature. If you think about it this makes perfect sense the only other known conventional way of increasing the CO_2 in the atmosphere is for the animals to exhale or the forests to burn. When the earth is a ball of ice, its normal state, there are few animals to respiration and not much forest to burn. As the earth warms, plants grow, animals proliferate and release CO_2. The only other way is to assume modern humankind existed, proliferated, burnt hydrocarbons every 100,000 years, then mysteriously dies out, and leaves no trace. Archaeologists have found traces of early man 139,000 years ago and he was a pretty basic cave dweller and the only hydro carbon he burnt was his fire wood or was when a piece of animal fat fell in the fire.

Other interesting discoveries can be made just by studying the earth's temperature. It would appear the coldest earth temperature was about -7 degrees colder than it is now. This happened many times and the earth's average temperature about -4 degrees colder than it is now. Both are ice age temperatures and imply the earth has been mostly a ball of ice except at the equatorial zones over the last 400,000 years.

It may not be important but with all that ice mass on the poles and in accordance with the laws of 'constant rotational inertia' of a body not subject to external forces, the earth would have turned much faster. Any child knows when spinning around on a swing or tire if you bring your legs in closer to your body you go faster much faster in fact. Anyone knows if you turn a roast faster on a spit it does not burn. So, if the earth turned faster because of the ice, it would also not heat up as much during the days. In addition, all that ice reflects the sun's rays. In effect cold makes ice and ice makes cold and so you have an ever-worsening situation of a runaway freezing earth and requires something very significant to warm it up again.

This is probably why the earth has remained so cold for so long. When it does get warm, thank God or nature or something, or we would not be here. When it gets warm, it gets green, it slows down and it rains and deserts are turned in to forests and the birds sing and the animals run rampant in the forest. Unfortunately, one animal, mankind, breeds himself insensibly and exhausts the earth's natural resources and upsets the balance of nature up to a point anyway. That point may be close at hand.

When the earth warms, it also produces clouds which reflect the sun's rays and so a warm earth has a natural built in over warming protection system unlike the runaway cooling system.

In 1987 when I first started hearing about all this greenhouse gas global warming in the newspapers my first reaction was this is not correct. I then started to research it and found one book 'The Green House Myth', written by an Australian in Tasmania named John Daley, which became my bible and inspiration to delve deep into the roots, which I did for twenty-one years in fact. Funny thing is in 1988 there was an international convention on global warming in Launceston, Tasmania, Tom's hometown. Not only was Tom not allowed to present a paper he was banned from the gate by guards. I met Tom ten years later in Melbourne; we had lunch together two days later before I took my family to visit America's beautiful wonderlands Yosemite, Yellow Stone and the Grand Tetons National Parks. Also, by coincidence, this is the same day National Geographic wrote to me and refused to print my article about what I am saying now.

The scientific method to determine the earth's relative temperature from the ice cores is largely to measure the thickness of the layer and measure the ratios between the two isotopes of oxygen O16 and O18. When there is a large difference in the ratio, for reasons I do not yet understand the planet is warm and wet. When it is low, the planet is cold. Also, a thicker ice layer is indicative of a colder period.

3.3 The Milankovitch Mechanism

In any event if you believe the Greenland Ice Core data which I think is very plausible all though I am sure these are errors somewhere there always is, it is very interesting to note there is very good correlation with predicted temperatures by the Serbian scientist Milutin Milankovitch. In the 1920s Milankovitch predicted the ultimate cause of climate changes lies in the regular alterations in the earth's orbit around the sun. The ultimate cause of such climatic change lies in regular alterations in the earth's orbit around the sun. The Serbian scientist Milutin Milankovitch first appreciated the significance of such orbital change in the 1920s. By building on his theories, scientists have established that every 95,800 years the earth's orbit changes from being roughly circular to elliptical. As this happens, the Northern Hemisphere develops greater seasonality, while the converse happens in the south. This sparks off the growth of northern ice sheets. When a circular orbit returns, the North-South contrasts in seasonality are reduced, global warming occurs and the ice sheets melt. Alterations in the earth's tilt during its orbit also have climatic implications. Every 41,000 years, the inclination of the earth changes from 21.39 to 24.96 degrees and back. As this angle increases, the seasons become more intense: hotter summers, colder winters. The earth also has a regular wobble on its axis of rotation much like a spinning top unless perfectly vertical it wobbles in a processional motion, which has its own cycle of 21,700 years. This wobble causes our seasons to not coincide with a perfect revolution around the sun. Approximately every 1,808 years on 20 March the earth is in a new Zodiac sign. We are now in the age of Aquarius where the Aquarius constellation on 20 March is now opposite the sun. This wobble is caused by the magnetic field of the sun. This influences the point on its orbit around the sun at which the earth it tilted with its Northern Hemisphere directed towards the sun. If this happens when the earth is relatively close to the sun, the winters will be short and warm; conversely, if the earth is relatively distant from the sun when tilted in this fashion, winters will be longer and colder.

If then the proximity of the earth to the sun combines with the worst tilt also made worse by another tilt caused by the wobble you will have the most extreme weather conditions. These conditions probably exist now with the southern hemisphere summer receiving its highest radiation. Temperatures in Australia are at record highs reaching 45°C (117°F) drying out the forests and creating massive fires. Sixteen hundred homes just destroyed and two hundred people recently died. Two years ago, almost thirty percent of country Victoria was raged by fires. My Rose River country retreat cedar chalet home spared but all 150 acres blacked. The green has slowly returned as new sprouts from the ground and most of the trees resume growth.

3.4 Solar Radiation Changes

There are other reasons for a change in energy reaching the earth. The largest one not allowed for by Milankovitch is the changes in the amount of radiation or heat we receive from the sun. During solar flares or when the sun experiences intense explosions which sends fiery plumes of hot 10,000-degree gases, hundreds of thousands of kilometres into space towards. The earth receives massive amounts of cosmic particles consisting of charged electrons and protons, neutrons and ultra violet light (UVL) and ultra-high frequency radiation.

This causes problems on earth with the radio and global position communication systems. This is one of the main reasons most intercontinental communication is through sub oceanic cable systems. Other problems include increased risk of sun cancer especially in Australia and Antarctica which is closer to the sun in their summers (Dec–March) and has occasional ozone depletion areas which would otherwise help neutralise some of the radiation.

The earth's magnetic field deflects the positively charged protons and helium ions one way and the negative charged electrons the other way. Many are deflected out into space. Those with a path more normal to the earth especially the much heavier charged particles like the protons and helium ions don't get deflected away as easy and end up hitting the earth. These with the neutral charged neutrons hits life forms on earth like tiny bullets which destroys cell's DNA and causes plant and animal mutations, cancer and aging. Stay out of the sun. Even golf courses according to my brother-in-law agronomist, Bill Collins, cannot escape the devastating effects as mutations in the grass greens cause a rough irregular green not good for the golfers and requires replacement every five years.

The northern lights in the northern hemisphere and southern lights in the south hemisphere are a usual effect of these charged particles in large numbers, swarming around a very much tighter radius at the poles bombarding each other and the air creating a light show.

While not yet proven it is my theory (Lincoln Magnetic Earth Theory) that the earth is positively charged as more of the heavier positively charges proton and alpha particles would get through the earth's atmosphere and not deflected away. The magnetic pole direction of the earth and the rotation of the earth are consistent with a positive charged earth. This theory also explains why lightning strikes the earth as the negative charged electrons are attracted to the positively charged earth.

From soil samples taken at equal and opposite distances from the mid-Atlantic ridge in the Atlantic Ocean the earth's magnetic field reverses every 300,000 years. Is the earth flipping up and down every 300–400 thousand years from some large reversing magnetic field, which could happen? Take a large magnet and put it next to a small magnet with an opposite pole and watch the small magnet flip over. Does this have something to do with 400,000 years of ice? Who knows? Since the earth wobbles or processes like a spinning top as it spins creating its own magnetic field. An external electromagnetic field particularly at right angles to the spin can cause this wobble to increase in amplitude if the frequency of the radiation matches the wobble frequency. This is called resonance which is when a driving fluctuating force matches the natural tendency of the object to vibrate or move like a pendulum on a string or a Childs swing. Under resonance conditions the object will move substantially. If the frequency of the ocean waves matches the natural roll response of a ship it can roll over.

A branch of physics called nuclear magnetic resonance spectroscopy uses this concept to identify organic compounds. Like the spinning earth in the sun's magnetic field, the spinning protons in the nucleus of an atom also wobble, and each atom and molecule has a unique characteristic magnetic dipole (like a magnet) and a precessional wobble. Small atomised amounts of a substance are placed in an external constant magnetic field. A fluctuating electromagnetic field at right angles causes the atoms wobble to increase substantially when the applied frequency matches the natural wobble frequency. When this happens, the atoms absorb energy which can be detected by the amount of absorption of the intensity of the applied magnetic waves. This same process is used in micro wave ovens to vibrate water molecules, and to activate concentrated water molecules in tumours which can be located when they give off their own radiation readjusting to the earth's magnetic field. So, if we were to get a large magnetic field resulting from somewhere in space at the frequency of one every 21,700 years, our earth's wobble and tilt would be substantial, possibly causing the earth to flip upside down as well if intense enough.

The earth's magnetic orientation could also change if the rotation changed direction or if, somehow, we get a periodic very large blast of electrons only that makes it through the earth's magnetic field.

3.5 The Sea Level

The sea level rises and falls as the earth's temperature falls and rises respectively. As the earth cools, the sea water contracts and becomes denser and sinks to lower levels in the ocean. This ensures adequate mixing of the waters of the oceans and helps give rise to Global circulation currents in the ocean. Cold temperature and polar waters sink and head either South or North towards the equator or to deeper water regions. In the case of the Atlantic Ocean cold waters move deep and south and warm tropical waters move north on the surface to replace the displaced cold water and this gives us the Gulf Stream. The Gulf Stream provides heat to South Florida and is why it is subtropical and has large thunder storms and rainfall and tropical plants thrive. A few hundred miles north of Miami the Gulf Stream heads more East and Florida loses its tropical climate and somewhere around Ft. Pierce where my mother lives is the last single standing most northerly coconut tree marking the end of the Northern tropics.

The Gulf Stream also brings warmth to Europe and especially Great Britain and Ireland. The warm waters heat the cold high-pressure air where tremendous evaporation occurs and then blows west-providing rain to most of Europe.

European sailing ships especially the Spanish ships from 1530 to 1716 controlled by the Holy Roman Catholic Empire which included most of Italy, France, Spain and Austria caught easterly trade winds and sailed to the Americas then caught the Gulf Stream and sailed north and then they caught the westerly winds and returned with the new world's riches back to Europe. It became apparent to me during my Spanish treasure searching I do with my son, Sean every year in the Florida Keys that the silver pieces of eight and golden doubloons are all stamped with the coat of arms of those countries. I had no previous idea the Hapsburgs of Austria used to run Europe then and finance the Spanish conquests of the Americas. It also explained to me how the trained Spanish Lipizzan white stallions were part of the Austrian heritage.

The Mediterranean seas warm waters spill on the top through the straits of Gibraltar as cold water moves in to replace it underneath. This process of global circulation of waters provides energy, oxygen, food and a flushing away of pollutants from bays and seas.

One anomaly of fresh water unlike any other substance on earth is that reaches maximum density at $4°C$ and then expands and becomes less dense.

The cold water then being lighter stays at the surface and freezes into ice. The ice layer not only insulates but acts as a greenhouse type glass and allows even in winter, sunlight and warmth to get to the water and lakebed below and ensure the animal life survives. Salt water gets more dense with the decrease in temperature at the poles and sinks to the deep oceans towards the equator, helping to create the circulation currents. The freezing point is depressed to -1.9 degree Celsius and because of these two factors it is difficult to freeze salt water and only happens at the surface at very low air temperatures at the poles and in winter only.

In general, the ocean expands with increased temperature and also to a lesser, amount melting of the ice at the poles and in huge glacial land masses in Greenland, Antarctica, Europe, South America and Alaska causes the sea level to rise.

The National Geographic fold out chart has some how predicted the rise and fall of the oceans along with the estimated temperatures and measures CO_2 concentration locked in the ice cores over the last 400,000 years. The chart, although the method not identified, estimates an average sea level of about 60 meters lower than it is now and about a 110-meter maximum lowering just before a major warming begins including at the end of the last glacial maximum 20,000 years ago. The sea level appears to have risen 105 meters or about 7mm per year in 15,000 years. It then appeared to remain at a constant level for the last 5000 years. The above levels appear to agree with historical data especially ancient underwater excavations of primitive shore dwellers 10,000 years ago in the Mediterranean Sea. It has also been confirmed from excavation of fresh water shell fossils in the Black Sea, which is now salt water. As the seas rose, they broke through the narrow separation of land between the Marmara Sea and the fresh water Black Sea now called the Bosporus Channel.

Other indicators of sea level are locked away as fossils in sedimentary rock and especially in limestone. Limestone or calcium carbonate is deposited in warm seawater. Because many of the sediments are so old "Geological Time" millions of years ago it is not clear if the sea level changed or the land elevation level changed from sediment weight or tectonic activity. Probably both things have happened and reversed and changed hundreds of thousands of times. In any event 2000–4000-feet thick limestone exists deposited from sea animals like oolites all over the tropical and subtropical regions of the earth. Most of Florida, Bahamas and the South Pacific Islands bed rock are limestone indicating the relative level of the sea at the time was much higher. Other deposits of marble around the world, which is metaphorised limestone like in Italy and Malaysia, include mountainous sediments now that were once under the ocean.

3.6 A Brief History of Global Warming

1980 to 2000	Earth's temperature increased 0.5 °C. Validation actual measurement.
1940 to 1980	Earth's temperature decreased 0.4°C. Validation Actual measurement.
1900 to 1940	Earth's temperature increased 0.4°C. Validation Actual measurement.
1886	Earths average temperature falls 4 degrees. Frost and freeze records in the Northern hemisphere. Reason Krakatoa volcanic eruption. Validation written history.

1815	The year with no summer, after the Indonesian volcano Mount Tembora erupted. Hundreds of thousands of people died in Indonesia and from the cold in the USA and Europe. Validation written history
1100 AD	Vikings leave Greenland as it has become too cold and no longer green.

Validation written history.

3000 BC	Great civilisations were built and man lived in large communities. Science, Art, Architecture, Mathematics, Astronomy, Philosophy, Literature and international trade flourished. The great stone buildings and monuments of Egypt and Mesopotamia appear.
9000–3000 BC	The earth gradually warmed to its present levels and mankind flourished in Asia and Europe he learned to make cloth, use horses, transport, make jewellery, bronze weapons, and lived in small agricultural communities.

Validation archaeological excavation the current earths warming period.

10,000 BC	The Younger Dryas period and start of the Holocene Period where the earth's temperature fell and a mini-ice age reappeared but for only a thousand years.

Mankind moved south and settled along the Mediterranean and rich river valleys of the Nile, Tigris and Euphrates rivers.

13000 BC   Late Glacial – Interstitial period where the earth warmed significantly and start of the agricultural phase of mankind where he learned to raise crops and animals, lived in larger communities and built mud brick homes.

Validation from Ice cores and archaeological findings in East Asia and Europe.

20,000 BC   Last Glacial – maximum when the earth started to warm and traces of cave men Neanderthal and Homosapien start to appear in Western Asia and Europe.
Validation Greenland Ice cores and archaeological findings.

40,000 BC   Start of another Glacial Period caused by the largest volcanic eruption ever known (Mt. Toba) which is now a massive large lake hundreds of miles across located in Sumatra, Indonesia. Volcanic dust in Greenland ice cores can be traced back to its origin at the Toba area. Palaeontologists have noticed a very small number and size of foraminifera sediments that have been recovered from drilling underground back to this area of time.

130,000 BC   Harsh cold glacial period where the earliest re-
                        mains of Homo sapiens have been found in
                        Ethiopia.

Validation ice cores and archaeological excavations.

400,000 BC   Evidently very warm as there are no earlier Green-
                        land ice deposits.

GEOLOGICAL TIME

33 to 5 million   Neogene Epoch – Large temperate forests would
                        engulf the planet and the carbon dioxide levels
                        would reduce from twice what it is now to only a
                        third of present levels.

66 to 28 million   Paleogene Epoch – The start of the many Ice
                        Ages caused after the large meteor described
                        below hit. The carbon levels would have in-
                        creased after the cooling and blocked sunlight
                        slowed down photosynthesis. The Epoch how-
                        ever culminated in a warm tropical planet that
                        saw the explosion of mammals, grasses and
                        flowering plants. Carbon dioxide levels would
                        range from 10 times current levels to only twice
                        what it is now. The higher oxygen levels would
                        allow mammals and birds to proliferate as they
                        are warm blooded and need oxygen to produce
                        internal warmth and also allows sudden energy

bursts necessary for flight and catching prey. This is unlike cold blooded reptiles that must hibernate in winter like snakes, iguanas and turtles and have to warm their body fluids by laying in the sun before they can metabolise energy.

200 to 65 million   Jurassic and Cretaceous Epoch – This time was characterised by large land and marine dinosaurs and large ferns and high carbon dioxide levels 5 times higher than now. This era culminated in a large mass extinction of the dinosaurs caused by a single meteor, 193 kilometres across that hit the Yucatán Peninsula, Mexico, putting billions of tons of radioactive iridium dust around the world called the KT boundary. It is believed this must have subsequently put the planet into a major freeze after the initial explosion and volcanoes caused exactly at the opposite side of the earth from the internal shock wave.

298 to 254 million   Permian Epoch – The largest mass extinction of all times called the Permian Triassic Mass Extinction that exterminated 95% of all life which included that in the oceans as discovered in sediments by palaeontologists. Did they freeze, burn up or die from radiation? No one knows. It is believed the largest meteor to hit the earth in Antarctica happened at this time and caused so many volcanos that poisoned the earth and the

oceans with hydrogen sulphide. High concentrations of carbon dioxide would follow as the cooling and blocking of sunlight would halt most of the world's photosynthesis.

358 to 330 million   Carboniferous Epoch The time when much coal found today is believed to have originated. The land was warm and humid and large ferns and simple trees proliferated near coastal swamps. Land vertebrae first appeared. Oxygen levels believed to be their highest ever.

541 to 489 million   Cambrian Epoch – The largest explosion of life on earth took place mainly with simple marine life forms in the oceans including sponges, worms, algae, trilobites and fungi. The carbon dioxide levels are believed to be the highest ever about 20 to 30 times todays levels.

4.0 THE CARBON ECONOMY – BRIEF HISTORY

4.1 The Industrial Revolution and Coal

The carbon economy started in England in the "second wave of advancement of mankind" in the early 1800s called the Industrial Revolution when mankind learned how to perfect steel making and make steam from burning coal. Before this mankind lived for 10,000 years as part of the first wave advancement of mankind called the Agricultural Revolution where people lived in smaller communities, raised crops and animals and relied on the not very abundant energy released from human and animal mussel energy supplied from grain which in turn was energy from the sun. This was also renewable energy and termed green today. It was a lot of hard work and people didn't live long then and literally physically worked themselves to death. The age of forty was considered old. Before 10,000 BC we were basically primitive hunters and gathers of seeds.

With the burning of coal in the Industrial Revolution huge quantities of energy was now available. Steam could power ships, trains, factories and even a few cars. The coal could be used to melt iron and make steel. Young adults left the farms that had been operated by their fathers and their fathers before then. The city was born and with it came different problems, overcrowding, pollution, rats, disease, open sewers, the plague, starvation and crime. We spent 150 years fixing the cities problems which gave birth to civil engineering, local governments, police, public works departments, water department, health departments, public schools, hospitals and psychological counselling and mental institutions. None of these existed before the advancement of the carbon economy. The nuclear family was born. This family largely consisted of a father, mother and children. The father worked mostly as a blue-collar worker. He had to be at work on time on the assembly line and the clock was then invented. The wife largely stayed home as a house wife until the 1940s when WWII plunged women into the work force. The children went to school. The grandparents were forced to stay on the farm and struggle for survival. Churches sprang up in every neighbourhood as a new social order with a new set of morals and ideals that put a lot of social pressure on the family to stay together.

Today the 'third wave advancement of mankind' or Super Industrialised Revolution starting in the 1960s spurned on by advance information technology (IT) including computers, mobile phones (cell phones for Americans) and now the internet has changed society and their values fundamentally. Most of the churches are left abandoned in Australia and sold as homes or restaurants. Many people remain single, more homosexuals are being born. The Nuclear family is almost dead.

Even companies and the work force have changed. A plumber, electrician, builder, brick layer, and other contract labour for buildings no longer need a secretary or answering service they can take their own calls with their mobiles or cell phones. Business, engineering and large corporate offices can now store reports, quotations, specifications and drawings and recall digital data easily modify and thereby reduce the necessity of a lot of office drafting and engineering staff. Factory workers are constantly being laid off and replaced by robotic devices. What ultimately is predicted is less and less traditional jobs that can be replaced or reduced by robots, mechanisation, and computers.

The burning of coal and the industrial revolution quickly spread across Europe and the Northern USA and later Japan. Burning of coal then resulted in excessive air pollution, buildings and streets were covered in black soot. Advancements in technology in the western world especially in coal fired steam power plants would somewhat reduce the emissions but this was expensive. Even today in China, the air and sky are mostly grey from the burning of coal with open coal burners for the last 200 years. Travelling to China as I have done since 1994 also revealed an absence of trees most had been cut down and burned as fuel in homes. With smog and carbon particles come many lung diseases including tuberculosis, cancer and emphysema.

4.2 Oil Development and Saving the Whales

Oil as we know it today was discovered and developed in 1854 in Pennsylvania, USA by George Bissett "Father of the oil businesses." Nobody knew what to do with it previously except as bitumen a highly concentrated viscous form to seal wooden ship planks or for brick mortar as had been doing for thousands of years. George one day looking at a tar pit in Pennsylvania had an idea it might be able to be treated and burned in oil lamps. The price of whale oil was getting more and more expensive as the Atlantic whale population was almost decimated from the use of their oil in lamps. Ten years later thanks to some heat processing and the invention by the Austrians of the curved shape glass lamp, gave rise to the kerosene lamp. Not only would the whales be saved but mankind and his amount of control and release of huge energy would revolutionise the planet like never before. The Industrial Revolution could now proceed at a much faster pace.

4.3 The Big Oil Companies

John D. Rockefeller of the United States started with a small oil refinery in 1880 and was quick to cash in on the demand for kerosene and developed Standard Oil which made him a billionaire soon owning the world's largest company. Gasoline at the time was a highly flammable dangerous liquid and was dumped in the rivers and streams. The first automobile was invented by Ferdinand Porsche and what would follow would soon take all of the gasoline that could be produced thus saving the environment. Interestingly to note oil was developed largely to save the whales and hence environment. Cars were developed as a tool for transportation that could utilise all the wasted gasoline or petrol that was destroying all the river and lake habitats and hence save the environment.

Standard Oil would eventually be broken up into seven separate companies; Exxon, Mobil, Chevron, Sohio, Amaco, Conoco, and Arco after the US government prosecuted them under violation of antitrust laws.

Shell Oil once the largest oil company but now quite small in comparison to Saudi Aramco, was started by a young Jewish boy named Marcus Samuel in London whose father made jewellery boxes from shells. This boy had the insight to see that the Dutch Far East Trading Company was providing most of the Europeans oil from Indonesia shipped around the bottom of Africa. At the same time the Suez Canal was nearly finished through Egypt providing quick access from the Indian Ocean through the Red Sea and through the Mediterranean to Europe. He did a deal with king Nasser to have the exclusive right to pass oil tankers through the canal and in return would pay a reasonable tariff to the king. The Dutch ships easily paid for the rite of passage and King Nasser, the Dutch and Marcus Samuel all profited handsomely. The shell jewellery shop and the Far Dutch Trading Company eventually became the Royal Dutch Shell Group. I always wondered why a Dutch company was headquarters in London. In

1892 Samuel broke Standard Oil grip on the worldwide Kerosene market. In 1862 Russians were burning American oil in their lamps. This would soon change after Russian Swede, Robert Noble, older brother of Alfred Nobel, inventor of dynamite and creator of the "Nobel Prize" was sent to Bacu by his brother Ludwig to find suitable walnut lumber for their riffle manufacturing industry. Instead, he discovered a small incompetent czar run oil industry that used to dig oil tar by hand and had some crude refinement laboratories. Using his brother Ludwig's money for the wood he instead bought a small refinery. The Nobels were in the oil business and Ludwig Nobel became "The Oil King of Bacu," within two years. The wealthy Jewish French family the Rothchild's would invest in a railway to bring Russian oil to Europe and also started their own oil company "Black Sea Petroleum Company." This company in 1912 would be bought out by the Royal Dutch Shell Group and the Rothschild's paid with substantial shares of Shell Oil.

In 1900 a Persian named Antoine Kitabgi travelled to London to find a capitalist to invest in and help develop their untapped oil industry. He met William D'Arcy who had previously immigrated to Australia as a solicitor forty years earlier and became rich in the gold-mine industry. D'Arcy formed an oil company called Anglo Persian Oil and hired George Reynolds to run things in Persia and nearly went bankrupt drilling for oil for five years until June 1904 when he hit a producer. Next, he needed a buyer and turned to the British Navy the Admiralty. The Admiralty arranged for Burma Oil a small business in Burma, selling oil to India, owned by Scotsmen to bail D'Arcy out. In return the Admiralty would buy oil for some of its small destroyers. Burma oil over the next ten years would struggle to survive until Winston Churchill switched Great Britain's Naval Fleet from coal to oil and bought most of the Anglo Persian oil company stock to ensure its survival. As a result, Britain won world war   as oil fired ships could gain speed more rapidly than coal fired ones. A company in Great Britain that distributed petroleum and owned by a German company named British Petroleum was taken over by the British after World War . This company was purchased from the British government by Anglo Persian Oil Company and changed its name to British Petroleum as we know it today.

The three major oil companies, Standard Oil, Shell and BP would battle for supremacy for a half a century. They would cycle between prosperous times and bankruptcy many times as general market forces of supply and demand would have it.

4.4 Automobiles and Saving the Environment

In the late 1800s and early 1990s the automobile was invented by Ferdinand Hearn Porsche. About the same time Ford in America was also developing automobiles. The purpose of the automobile was to find a use for gasoline, a highly volatile and dangerous by product of refining oil for kerosene lamps. Before the automobile, gasoline was dumped into the rivers and streams causing extensive destruction of the environment. So, the automobile actually saved the precious rivers and coastal estuary environments.

Mankind's business ethnic agreed really started in the early 1930s and caused America's total dependence on the automobile and oil. This was kick started when Standard Oil, General Rubber, and General Motors formed a consortium of sorts and bought up and scraped all of the mass transit street cars and train systems operating then in every major US city. Evidently there were no antitrust laws to prevent companies of different services collaborating together to their mutual benefit and the eventual dependence on oil and now possibly the destruction of the world's Western economies.

By the late 1940s and the early 1950s gasoline was king as tens of millions of cars filled the roads. There was an excess of kerosene now and this lasted until the jet engine saw commercial development in the 1960s and kerosene was renamed jet fuel. I remember my father worked for Aerodex, a large South Florida airplane engine service company who was laid off in 1967 as jet engines replaced propeller driving gasoline engines.

4.5 World War I and II

During World War I Great Britain had their steam driven war ships converted to oil which could more quickly gain speed. As a result, she defeated the German navy. Prior to World War II Japan had built up a large army and navy using oil supplied from the USA, the world's largest producer of oil at the time. The USA was worried about Japan's worldly aggression especially against China and cut off their oil and therefore their food and livelihood. The Japanese quickly retaliated and attacked Pearl Harbour and as a result bringing the "sleepy giant" into the war as Japanese admiral Yokamoto called the USA. The WW was fought about oil and won because the USA had lots of it. The German desert foxes General Rommel and his tank army was defeated by the allies in Africa largely because they ran out of fuel. It was a sheer coincidence that under these abandoned tanks deep in the sediment was some of the largest oil finds in the world not yet discovered.

Later major discoveries of oil fields were found in the Middle east including, Kuwait, Saudi Arabia, UAE, Qatar, Oman, Iraq, Iran and Libya. A few lucky business entrepreneurs in the 1940s and '50s struck up business deals with the leaders of these countries to develop their oil and literally became billionaires overnight.

4.6 OPEC Oil Shocks of the '70s

Eventually most of the oil would be discovered in third world countries in the Middle East and also Mexico and South America. These countries would form an oil cartel called Organisation of Petroleum Exporting

Countries (OPEC) and would raise and control the price of oil and nearly destroy the worlds western economies like they did in the 1970s, and they are trying to do now in 2008, when oil went from six US dollars to thirty-nine dollars per barrel in five short years. What resulted was high double-digit inflation as the price of everything sky rocketed associated with the massive cost of oil, petrol chemical products and transportation of everything. I remember living in South Florida at the time; my family's house bought for $10,000 in 1950 was sold for $70,000 in 1974. I remember the long gasoline lines associated with the gasoline shortage and the rationing system that only allowed either odd or even number plates on a given day. I remember thousands of stranded vehicles on the side of the road. This woke something up in me about the instability of our economic system and the inability of the government to quickly cope with it.

I had just graduated from Florida Atlantic University with a degree in Ocean Engineering. Employment was readily found by me in the offshore oil drilling, offshore pipeline construction and consulting engineering of oil and gas facilities for the next forty years.

Many of the world's countries started switching to alternative forms of energy mainly natural gas and nuclear to avoid the high cost of oil. Margaret Thatcher of England had many nuclear plants installed and today nuclear energy in Great Britain supplies about 35% of their electrical power needs. France the world's leader in nuclear power technology currently runs over 80% of their power on nuclear energy. Russia and the USA started to convert to nuclear but fears of a nuclear disaster slowed down the momentum after the accidental release of radioactive steam at High Island in the USA and the most feared disaster of all a nuclear fuel meltdown, which happened in Chernobyl in Russia. Evidently the USA and Russia had developed enough nuclear weapons to destroy the earth ten times over but were not prepared to figure out how to use the energy for peaceful purposes. Thank God for Western Germany affluence, which made communist Germany and the rest of the Soviet Union take notice that capitalism was a better economic system. Also, thanks to Gorbachev, the leader of the USSR, and drinking buddy with Ronald Regan (Raygun Ronald) did not fight it when member counties divided to succeed from the Soviet Union causing its collapse in 1989.

Japan switched to natural gas which was and still is relatively inexpensive about 20 to 30 percent the price of diesel or gasoline. Japan not having any natural oil or gas reserves on their own island would embark on a new business strategy for securing their energy needs. They would finance the cost of developing gas fields in Australia, Indonesia and Saudi Arabia and build liquid natural gas LNG plants and LNG transfer ships. In return the gas in the ground was virtually theirs and the price was fixed often for up to twenty years. Unfortunately, when they would lock into a fixed gas price for twenty years it did not help them much when the gas price dropped significantly as it did in early 1980s. This move actually cost them much more for this energy than they would have paid throughout the '80s and '90s as energy was cheap. This combined with the excellent quality products they produced including cars, cameras, stereos and watches which lasted for twenty years would eventually put their economy in recession for ten years from 1995 to 2005.

In the '70s and '80s, natural gas was largely just burned at the well head production platforms. I remember travelling at night on crew vessels to some distinct place off the coast of Louisiana or Texas and observed thousands of offshore platforms lit up from the gas flaring. The vessel captains didn't have DGPS back then for navigation so they would pull up to a platform and read the platform identification including the production block, name and well number, much like a taxi driver reading a street number to know where he was. A chart with all the well locations would soon show you where you were and which direction you needed to head. From a distance it looked like hundreds of yellow stars on the horizon. In offshore Western Australia on one of the first oil production sites at a place called Varnos Island, flaring was not allowed as it was believed to disrupt the breeding habitats of the turtles. I don't know why but I always chuckle when I think of two turtles making love at night floating on the surface of the ocean, silhouetted with an orange flare in the background. If it was not romantic to them at least they might have better detected an offshore crew or supply boat running into them. But that is Australia for you. Later, to the betterment of mankind, flaring was outlawed almost universally as part of a Waste Reduction Policy, the only good thing to come out of the United Nations Kyoto Protocol agreement. Today Australia has some of the world's largest gas supplies and has developed two twenty billion dollar and two more planned plants to liquefy it (LNG) and export it worldwide. I worked as an engineer in my early days in Australia on one of the plants off and on for ten years. The Varnos Island gas plant blew up in 2008 from failed internally corroded piping creating a large fire and reducing a third of the gas supply to Western Australia.

4.7 The Oil Crash of the '80s and '90s

In the early 1980s the price of oil fell to USD ten dollars per barrel as an oversupply developed. In the late seventies and early 80s oil companies found and developed more oil fields around the world. New power plants built were running on nuclear or coal fuel. In France the world's leader in nuclear technology was supplying 80% of their power needs from nuclear fuel.

OPEC saw less oil demand from the competing energies so decided to corner the energy market and over produced to drive the oil price down and competition out. It worked up to a point. In Houston in 1985 a thriving boom town oil and gas town had turned into almost a ghost town. I remember living in Australia being mostly out of work and went to Houston to sell my house. You could not sell a house. People with no jobs fled leaving the key in the door for the mortgage company to absorb the loss. Twenty-five storey beautiful gold and blue modern gas plated buildings were empty everywhere. These buildings used to house offices of hundreds of oil service companies and one now had a vacuum cleaner service centre on the ground floor as the only tenant. I couldn't sell my house but rented it out. I returned to Australia as a contract consultant mechanical engineer and worked in the mining and power generating fields until the early 1990s.

Through the 1980s cheap oil was a disaster for the oil business but the rest of the worlds industries thrived. By the mid to late 1980s inflation set in especially in Australia. The price of homes was doubling in value every five to seven years and the interest rates also went up, way up. Many young Australian families borrowed hundreds of thousands of dollars from banks to buy a home they could not afford on the assumption it would increase in value. In 1987 the day of reckoning came as there were major stock market collapses globally. People switched their investments to homes. Two years later the interest rates hit 18% and homeowners now especially those on variable rates, which most Australians have, had to pay for the financial excesses of the failed business and irresponsible bank lending. By the late '80s and early '90s individuals, businesses and lending institutions were all going bankrupt. What many people don't understand when you buy a home and borrow money especially on a variable rate, the interest you pay will be financing all the money that was lost by greedy, dishonest, incompetent and irresponsible businesses and lending institutions.

After the recession in the early 1990s, the price of oil stabilised at about USD $20 per barrel. In 1992 president Clinton was elected and the world experienced the longest period of steady growth of all times. President Clinton, probably one of the greatest presidents of all time as he balanced the budget and paid back the trillions of debt the country had inherited from the Reagan and George Bush senior presidential times. President Clinton was viewed by the world as a friendly, peaceful and benevolent man. As a result, there was little or no terrorism and consequently no curtailing of the oil supplies. Mr. Clinton also loaned third world countries like Mexico billions to stabilise their countries, which they paid back with interest. Mr. Clinton oversaw the peaceful demise of the Soviet Union and the end of the cold war. Mr. Clinton also helped organise an international defence force through NATO to stop the war and ethnic cleansing of Muslims in Bosnia and Croatia. USA forces assisted in the air and not one American life was lost.

Okay, he couldn't keep his pants on, but what great man can. It was a well-known fact that Franklin Roosevelt had a few affairs not to mention Benjamin Franklin literally the "father of our country" that had a rumoured thirty affairs and many illegitimate children. Hilary Clinton once in an interview when asked about her husbands alleged womanising replied 'Bill was a hard dog to keep on the porch." Anyone ever owning a male hound dog knows exactly what she is saying. I had the most beautiful male Walker Hound a Fox hound named "Duke" when living in Florida, Nova Scotia, Alabama, and Louisiana in the 1970s. Ole duke would get up sniff the air, smell a female in heat miles away and would disappear for a few weeks only to come home all beat up and bloody. I am sure Bill was not that bad.

4.8 The Kyoto Protocol Blunder

It's hard to believe in the twenty-first century since we are all so educated, aware of histories past blunders and intelligence is now global with the aid of the idiot box, the television, that the world's governments could team up and create this Kyoto Protocol blunder.

Let's face it we are a pretty ignorant and gullible race. We used to burn women as witches and believed the world was flat and ships would fall off the edge. The USA civil war was about economics not slavery, which was the human rights excuse they could sell. Much like Japan in WWII tried to sell to the Asians their invasion of their lands was about the removal of the white people from Asia.

We need to be more careful and not accept on face value what scientists and especially what the government tell us. The fact is most humans and governments are motivated by personnel greed, power and sex. As a result of global prosperity, from low energy prices in the early '80s and the great benevolent leadership of Bill Clinton between 1992 and 2000 the USA and the world had the best slow steady economic growth of all time. There was little wrong with the world and a few underpaid not credible scientists decided to get together and create problems by creating the "Global Warming Myth." From the mid-1980s to the late '90s, everyone was jumping on the bandwagon and the public wanted to know and so hundreds of millions of dollars of grant money was made available. During this time there was over 165 technical publications made predicting doom and gloom on how global warming and increased CO_2 was going to destroy our coral reefs, flood all the worlds low islands, destroy the polar bears habitats, increase the ferocity and frequency of hurricanes destroy the fishing industry, destroy the crops, cause excessive rains and also droughts at the same time if you can figure that one out. Nothing was safe Global warming became the blame for any adverse weather including snow storms.

Governments were enacting legislation making it illegal to cut a single small tree even in a million-acre forest or even your trees at a suburban home threatening to crash on home roofs or roots ripping up their sidewalks. Children were coming home from school depressed and told by their teachers the world was going to flood soon. No one at the time even questioned whether the warming was actually taking place or even if CO_2 played a roll. It was just assumed by most of the world's cities population this was happening as most of what the suburbanites see is cars, industry smoke and steam stacks. Governments got on this bandwagon big time. In 1992 the United Nations member countries accepted without proof excess CO_2 and global warming. This resulted from previous world meetings where global warming was falsely assumed. These meetings included:

General Assembly resolution 44/228 of 22 December 1989 on the United Nations Conference on Environment and Development and resolutions 43/53 of 6th December 1988, 44/207 of December 1989, 45/212 of 21st December 1990 and 46/169 of 19 December 1991 on protection of global climate for present and future generations of mankind.

Recalling also the provisions of General Assembly resolution 44/206 of 22nd December 1989 on the possible adverse effects of sea-level rise on islands and coastal areas.

Recalling further the Vienna convention for the protection of the Ozone layer, 1985 and the Montréal Protocol on substances that deplete the ozone layer 1987.

Also, the ministerial Declarations of the second world climate conference adopted on 5th November 1990.

The green house myth continued to snowball and gain momentum and culminated in Kyoto, Japan on 1–10 December 1997 Agenda item 5 Kyoto Protocol to the United Nations Framework convention on climate change. By now the whole world was steaming in their juices.

During this time United Nations delegates from the forty member countries met in Japan and largely stayed drunk in Japanese Geisha houses drinking sake and spirits. The next day with all the hangovers, even if they showed up on time and didn't fall asleep, they were all very agreeable and easily lead astray except Australia. You can't pull the wool over the eyes of

Australians and they know how to drink and still function the next day. Australian men as a rule don't chase women and prefer to drink with their mates.

When I think of the Kyoto conference, I can't help but relate to the Commonwealth Head of Government Meetings CHOGM held each year in a British Commonwealth country. Since most CHOGM delegates are black from predominately black populated countries in Africa and in the Caribbean and love to drink and party, the Australians refer to CHOGM as Coons Holiday on Government Money. I say this jokingly as I am not a prejudice person and I can and do say nigger, Jew, Coon, Spic, Limey, Wop, Wog, Rag head, Kraut, Pome, Wasp, Coolie, and Yank. And I say so in a jokingly respectful way. In Australia we have no racism and always joke about each other's nationality. I hope one day in America I can see a black man shake the hands with a white man and call him honky and the white man calls him nigger and both laugh without public outrage. America will then grow up and truly be free of racism.

Twenty-seven Articles written on twenty-one pages set in place strict guidelines for the member countries in reducing greenhouse gas emissions and establishing methods for removal of these gases.

In general, all member countries were to reduce their net carbon dioxide emissions 5% below each country 1990 levels between 2008 and 2012. Carbon credits allowed countries to produce more if they could prove they reduced emissions in other countries by selling better technology or by removal of CO_2 from increasing removal sinks like associated with reducing deforestation.

Carbon credits are a form of currency used by rich countries that have signed the Kyoto Protocol, a global pact to combat pollution. Every signatory has agreed to cap their emissions of greenhouse gases such as carbon dioxide and methane, believed to increase the atmosphere's ability to trap infrared energy and thus affect the climate. The cap varies from country to country; based on this, each is allocated a certain amount of carbon credits. One credit represents the right to emit one tonne of greenhouse gases. The credits create a market for reducing greenhouse emissions by setting a monetary value to the cost of polluting the air. There are two types of carbon credits; offset credits and reduction credits. The former promoted clean forms of energy production such as wind or solar; the latter consists of the collection and storage of carbon from the atmosphere, such as through reforestation. Developing countries such as China and India do not have to comply with limits on their emissions at present. They were excluded as they were not the main culprits of emissions before the protocol was set up in December 1997 in Kyoto. As for the rich countries - many of which fall short of their targets for cutting emissions—carbon credits buy them time to reach their targets.

Australia objected coldly as they are a net exporter of energy especially coal and it would adversely affect their economy. By 2008, most member countries admitted they cannot afford to comply and followed Australia's lead. No one questioned that maybe if there was warming and an increase in CO_2 what the advantages there may be like deserts and permafrost areas turning into arable land and food production being higher. Australia was however one of the first countries to adopt a "Waste Minimising Policy," which among other things outlawed the flaring or burning of well head methane from oil production rigs.

Live coral reefs are as important to our environment as botanical gardens are to our land. They look nice but that is about it. Besides coral polyps are platonic and float in the ocean and will grow whenever the right conditions of clean water, sunlight, warmth and dissolved oxygen combine to their liking. We have limestone 4000-foot thick with coral rock abundantly inside which clearly indicates as the seas warm and rise the corals grow upwards.

5.0 THE CHAOTIC CARBON CRISIS IN 2008

5.1 Introduction to the Economic Meltdown

At the time of starting to write this book in March 2008, oil is USD 130 per barrel. Middle east tensions and Muslim fundamentalists terrorism is at an all-time high a housing price collapse in the USA which has spread overseas called the "Sub-prime crisis," has left many lending institutions bankrupt including major international banks, interest rates are unstable and world money leaders don't know whether to raise them or lower them. Inflation is sky rocketing. The US dollar has fallen through the floor and the unemployment rate is soaring, the stock market globally has plunged if not crashed and American news reporters are asking each other if their country is in a recession or not. The answer is it is possibly the start of another major depression but let us hope not.

How did all this happen? What happened to all the economic prosperity and growth and low oil prices and friendly Islamic and South American countries we had from 1992 to 2000 when Bill Clinton was president. Those were good times, but unfortunately, we became too reliant on the importation of cheap oil to run our economies.

5.2 The Bush Administrative and Iraq: The First Carbon Conspiracy

The answer to the question above is obvious; George W. Bush, son of George Bush Senior, both oil tycoons and war mongrels was elected to the presidency.

It didn't take the Muslim fundamentals long to vent their anger when that fateful day on September 11, 2001 when the New York Trade Center Twin Towers and Washington Pentagon were bombed when jetliners loaded with jet fuel crashed into them.

After a couple of years of bombing and invading Afghanistan, which didn't yield Osama Bin Laden, the state department decided to invade Iraq for reasons only the Bush family know about. The reason given was the Saddam Hussein was developing weapons of mass destruction and under United Nations Security guidelines he could be stopped. Problem was after the invasion and even now no weapons of mass destruction could be found.

The invasion of Iraq was and still is about securing oil for the West, the "First Carbon Conspiracy."

Iraq the second largest oil producer in the world had not been allowed to produce oil for profit since 1995 when George Bush Senior invaded Iraq. It is common knowledge that the USA had put Saddam in power to help fight Iran. According to an Iraqi neighbour of mine whom also is an executive for Mobil Oil whom lived in Iraq during desert storm, Saddam conspired with Saudi Arabia to invade Kuwait and divide up their oil fields. USA State Department officials were asked what would be the USA position if Iraq invaded Kuwait. This was relayed to George Bush Senior whom replied the USA would look at this as a domestic dispute and not intervene. They did intervene however a few days later and we had Desert Storm I.

So, Iraq for twenty years has not been producing any appreciable oil for world consumption.

5.3 Middle East and OPEC- The Second Carbon Conspiracy

The other Middle East countries, mainly Saudi Arabia and Iran, have also cut production in retaliation for America's invasion of their neighbour. This anti-American sentiment has spread to most of the OPEC nations including, Venezuela and Brazil and general cutting back on their oil supplies. This is the "Second Carbon Conspiracy," the cutting back on oil supplies by the OPEC and middle east countries for retaliation against the USA and other western countries that supported the invasion of Iraq.

5.4 China and India Feeding the Commodity Bubbles

China and India with about 2 billion people and two fifths of the world's population seemed to come out of the closet in 2000 and admit consumer capitalism is not a bad idea after all after about fifty years of isolation policies. What has made these countries want to adopt western consumerism and capitalism? Two things, firstly the breakup of the Soviet Union in which both countries had ties, secondly capitalism has won out over communism as more economical, and there is a lot more personal freedom. What had spurned the public awareness and interest to emulate the west in dress, habits and consumerism, one thing the television. The television is the 'dawn of intelligence of mankind' if you watch it and hear it you will always remember it much better than if you just read it. Intelligence is then not always a good thing.

I have travelled to both countries extensively since 1994 doing business development for my Oil and Gas Engineering and Construction Company.

Travelling to China in mid-1990 was to enter a predominantly communist country with a militant government complete with red arm bands and, green commander hats and with red stars. Many of the rural areas I had to travel to assess project constraints had never seen a white man much less speak English. Most people were very friendly though and I was only kidnapped once in Shanghai by one woman and three men when I got into the wrong taxi. They were disappointed I had no money, but I promised each of them five dollars if they took me safely to my hotel, which they did. I remember my first trip in 1993 to visit CNOC (Chinese National Oil Company) they had a difficult subsea pipeline to install and wanted someone to tell them how to trench it. I remember sitting by myself on one side of a very long conference table. Only the other side was twelve Chinese military men dressed in traditional green with red star hats and uniforms. The top guy, also the largest and maybe the oldest sat directly across from me. I felt like sponge bob square pants at a bar with angry tattooed bikies. No one spoke English except this young Chinese girl who acted as an interpreter. I left and didn't hear from them for four years until they got into trouble and couldn't trench their pipeline. Their contractor a large Italian English consortium called EMC called me to Shanghai. The same big Chinese military bloke was there and recognised me. After twenty minutes of my sales speech the big military bloke stopped me and said something loud to me in Chinese, which someone interpreted to me "can you trench this pipeline," I replied jokingly "of course but you have to give me the soil." After a few seconds of silence there was laughter and someone relayed my joke and the big guy laughed as well. We were in and got the job.

Everybody road bicycles in those days in droves with constant slow speed collisions. I remember the landscape was treeless everywhere where and endless smog no matter what city you visited in China. There were many young intelligent twenty something English speaking college students about. Evidently, China enacted a one-child policy twenty years before. Before it became law, they all had loads of children. I go back to China now in 2008, and beautiful office buildings and expressways and cars are everywhere. The people still drive and crash into each other like they did with the bicycles. Trees are being planted but the smog is still there but a little less. No more armed Chinese red guards all around.

The Chinese economic miracles going from a third world communist country to a thriving rural industrialised capitalist country in fifteen years requires a major rewrite of all the "Isms" I was taught in school and economic philosophies of the past century. Almost all cloth and clothes, tools and toys are made in China nowadays. Now this has not been without problems, the Chinese by nature have a relentless drive to work hard, work cheaper and produce fast. Much of the time however quality and ethics in copyright and product piracy take a back seat to a quick production.

The Chinese are building roads, dams, steel mills, powerlines and large buildings at a phenomenal rate. One power plant a week is being commissioned. All this growth is creating a huge demand for commodities including, copper, iron ore, coal, LNG, and oil. This demand helps take all the excess commodities off the market and subsequent drives the prices high. Australia a mainly commodities exporting country is busting at the seams to mine, load and export mineral commodities to China. Every engineering mining and construction firm is flat out. Unemployment hardly exists and there is a wage explosion, shortage of skilled labour and high inflation especially the overinflated housing price bubbles, combined with the high interest rates has boosted its currency 50% to the US dollar. Australia has become the most expensive place in the world to live. Australians conventional labour governments' mentality prohibits cheap immigrant labour. This is changing now as cheap labour is being imported and disguised as political refugees and the spouses of fee-paying university students.

Now that the Chinese workers have some money they travel. They travel a lot around Asia and now if you can find a hotel room in Singapore or Thailand you will pay double for what you paid two years ago. I see hundreds of Chinese now travelling and many pile in a lift or elevator for the first time, and do not know what button to push.

The Indians are good people and speak fluent English, which is a major advantage. India is not growing as fast as China as they have a constipating bureaucracy they inherited from the British. No one is allowed to make a decision and everyone tells their boss what he wants to hear, not the truth. The Indians also not like the Chinese do not value their time much. One night in New Delhi instead of taking me to a one hundred dollar a night hotel, which used to be twenty dollars a night four years ago my agent drove around for a few hours and knocked on doors of guesthouses at one in the morning to save $50. I ended up sleeping in a room with four people on the floor outside my door. In the last ten years India's hotels have become too expensive, the homeless have been moved out of the city, inflation is excessive, real estate in Bombay is more expensive than Australia, more cars on the road, more young people not wearing conventional dress and also watching Western TV and movies.

The Indians have opened up their country to investment and foreign companies. When I first travelled to India in 1994, I was shocked I saw millions of homeless people living and sleeping on the streets of Bombay. Beggars who had their arms cut off by their parents when they were young could easily get more money. I had to travel by train to remote towns as there was no air travel then. On one overnight trip I was delighted my agent arranged a sleeper car. To my and especially their disappointment I had to share the car with a family, my worst fears were confirmed when I smelled something and rolled over to see the grandmother squatting 3 feet away from me, taking a dump into a bucket. Upon entering and leaving India, I was questioned and sometimes frisked and not made to feel welcome by immigration and custom authorities.

In 2008, we were doing a project in Paradip. From Mumbai or Delhi, you have to then drive to a small town then drive to Paradip, a four-hour drive. The last two hours you have to pass twenty thousand trucks or lorry's as they call them, loaded up with iron one going to the port and then by ship to China. The lorry drivers stay in line for about two weeks slowly inching their trucks forward. They cook and eat on the side of the road and take turns sleeping with the second driver. After dumping their load, they speed back away from the port in the opposite lane that you are in trying to pass the other nineteen thousand nine hundred and ninety-nine trucks. There are many collisions and many fatal ones as a result.

Everybody is flying everywhere by mid 2008 now and all the airports are congested. I was supposed to fly from Bhubaneswar to Mumbai at 1400. The flight was delayed until 1830 and while midair Mumbai control tower asked the plane to fly slowly and delay landing another hour. All this for a two-hour flight.

India's oil companies are expanding rapidly. Some like ONGC on the west coast have their own oil fields. The other two oil companies IOCL and HOCL import oil from the Middle East, refine, and distribute. No one trusts Indian companies including Indian Oil Companies and especially other Indian Companies. The amount of lying, cheating and extortion is unbelievable. Many Western Companies including my own have lost money and have pulled out of India or at least working for Indian Companies without explicit bank guarantees.

The rapid expansion of China and India has removed all the excess commodities off the market creating a shortfall and rising prices and inflation. In late 2008, it appears inflation is slowing down. Chinese and Indian stock markets have crashed and growth for China has slowed down to a single digit the first time in fifteen years. The USA, despite all the free trade stocks and NAFTA alliances of the previous years has managed to put a 30% tariff on all Chinese imported goods. Inflation has also taken its toll in India and China as the price labour and real estate had soared. Over the years, public sentiment against China for producing dangerous toys including those with lead paint and violation of copyright laws on almost all products has prompted governments to stem the uncontrolled inflow of Chinese goods to Western countries.

This anti-China and other countries sentiment is having a significant effect and will have a larger one soon. A new financial trading alliance called **BRIC** comprised by Brazil, Russia, India, and China which has almost half the world's population are forming strong trade alliances to reduce their dependence on the West. Soon they will produce all their own oil, food, cars, ships, military equipment and thanks to Air Bus opening plants in China, air planes.

5.5 OPEC – Creates Havoc again "Second Wake-up Call"

By December 2008, the economic leaders of the USA have acknowledged a full-blown worldwide recession. Thirty international banks have gone bankrupt and 400,000 Americans are losing their jobs every month.

The automobile companies have shed 500,000 jobs in Detroit as Americans quit buying gasoline guzzling cars. Why did American leaders go back to sleep after the wake-up call on expensive oil in the 1970s, couldn't the most powerful nation in the world understand its dependence on foreign oil and the 600 billion it pays each year to import it would destroy their economy. Could they have not helped reduce dependence by building electric trams and trains like existed in the 1930s.

Small oil companies and even moderate size ones have by now invested hundreds of billions in developing new fields to cash in on the oil profit wind fall. Alternate energy companies like one of mine, OES CNG have invested in systems to power and fill cars on natural gas. Other companies have invested in wind turbines and electrical cars.

OPEC leaders and especially the Saudis which own about 8% of America and whom have watched their wealth erode from the share price collapse have another trump card to play. They play the same card they did in the 1980s and start to increase production again to lower the price of oil, bankrupt all the new oil and alternate energy plans and make the Western world dependent on them again. This is the second wakeup call now twenty-six years later. Are we going to go back to sleep or wake up, pull our finger out of our ass and look after our financial future? If OPEC were American companies this would be called creating a monopoly and what they are doing would be illegal and they would be stopped. This is a good time for NATO and SEATO and the United Nations to get together to control OPEC.

The east and west have the persuasive power. Should we not have an OPIC "Organisation of Petroleum Importing Countries." Russia, a wild card for about seven months of 2008 looked at being one of the wealthiest countries in the world with the vast but remote and expensive to produce Siberian Oil reserves worth over 100 dollars a barrel. Now at 40 dollars per barrel they probably will have to join the ranks of the other consumers or else conspire with OPEC to raise prices. China a major importer would support OPIC and swing the worlds military might with the USA and Europe to the OPIC side.

5.6 End of 2008 – End of Capitalism as We Knew It

By the end of 2008 a Global recession firmly is in place. The high oil prices that caused the massive inflationary bubbles popped along with the housing, commodity and share price bubbles.

For a brief period, it looked like the oil companies and alternate energy companies would prosper until OPEC wrenched the oil valves wide open crashing the price of oil and taking with it a lot of the alternate energy hope and progress we so dearly need.

Thirty international banks have gone bankrupt and Americans are losing their jobs at a rate of 400,000 per month. Detroit has shed 500,000 auto workers. The US government is working on a plan to subsidise General Motors, Ford and Chrysler with USD 14 billion to minimise further layoffs. This plan was hinged on getting the AWA (Automobile Workers Union), to accept salary cuts, which they refused and now has costs doubts on the future of automobile manufacturing in the USA.

Financial companies including Bank of America, Merrill Lynch Citigroup and others have announced more than 250,000 job cuts in 2008.

The bailing out of private firms like Bear Sterns, Freddie Mac, Fannie May and American International Group by the US government marks the end of capitalism as we have known it. It was believed communism lost out to capitalism in 1990 with the demise of the USSR (Soviets) and now eighteen years later capitalism fails on a worldwide scale.

Things looked good for Russia to emerge as a very rich country and another world power until late 2008 where upon OPEC increased production dropping the price of oil from USD $150 per barrel to USD $40 thereby making expensive Siberian oil no longer economically viable.

Americans having seen their wealth erode first in their homes second in their stock shares and thirdly in their investment funds and loan institutions. For the first time in history with negative interest rates, Americans are so scared they are paying the US government to safely hold their money. Not really a wise move since it was the US Government that helped start all this mess.

In mid-2008, China economists did not think the economic crisis and large USA import tariffs would affect them as they had a very global export market. In November 2008 China had the first decline 2.2% in exports since June 2001 and a sharp reversal from the 19% gain in October and a nearly 20% rise in 2007, imports also fell 18% in November including 17.3% for crude oil imports. Chinese producers of, low-end goods such as toys and textiles had already been struggling for over a year as a result of the USA 30% tariff and by late 2008 the economy has deteriorated sharply and sales of high-end machinery and electronics has declined substantially as well.

The steep drop in the importation of oil by China has helped OPEC create a global oversupply of oil causing prices to tumble from US $140 per barrel.

China's economic growth is forecasted to slow to around 7.5% next year the lowest since 1999. In 2008, growth was 9% but the first-time growth has been less than a double-digit gain in five years.

Despite the advantage of Japanese green fuel-efficient hybrid and CNG alternate fuel automobiles which was believed to be in such big demand in the USA that they would escape the financial crisis, it is too late. The Americans and Europeans no longer have any money. The crash in the oil prices also puts green concept automobiles like hybrids and CNG cars back in the storeroom.

Overseas demand for machinery from Japanese has plunged 44% the largest ever. Mazda motors intends to slash production by 100,000 vehicles. Japanese prime minister Taro Aso claimed "this is a once in a century global slump, Japan isn't going to escape from this big tsunami but we can respond to it in a way that minimises the damage."

The collapse of Detroit is also acting negatively on Japanese auto makers as they rely on their US counter parts for parts and sales.

By the end of December, 2008, the Bloomberg CNBC and *The Wall Street Journal, Time,* and *Newsweek* are all claiming an unprecedented situation and possible global depression.

Everyone is turning to Obama the first black and first previously Muslim new president elect of the USA to solve the world's biggest economic problem. The last time a nation was in complete despair and almost universally elected a major world leader not representing the majority and fundamentally different from the electorate was in the 1930s his name was Adolf Hitler.

I doubt there is any other comparison between the two. I like Obama especially as compared to John McCain which represent the same old same old Republican excessive government spending on the "war machine," I am glad he has appointed Hillary Clinton as secretary of state as I know Bill Clinton "Super Bill" will now be in the background to save the economic world the second time.

Osama Bin Laden whom attempted to wreck the American economy in September 2008 has been successful within seven years he has also wrecked the world's economies. Let's hope he is found and brought to justice. It's the first round in the New Year Osama versus Obama and with Obama yes Chelsea Clinton we now have your mama. At the update of this book in 2021, Osama was found in Pakistan and taken out by the Navy Seals at the direction of Obama and Hillary Clinton.

6.0 A NEW WORLD ORDER

The people of the world need to and will unite and force their governments to change and embrace policies that reduce our populations, resource exploitation and dependence on oil, increase our renewable energy utilisation and recycling of goods.

A new Fourth Wave Advancement of mankind called the Green efficient and renewable Industrialised Revolution will inevitably take place.

6.1 Reduction in Populations and a New Economic System

The world's resources can no longer supply the demand of the populations. We need to decrease our populations somehow. China did it when they enacted a one child policy law in the seventies. It helped their economy emerge very strong. A new world order and an economic system that is a compromise between unregulated capitalism and communism need to take place. The Australian, Netherlands, Swedish, Singapore, and Swiss economies are possibly good models.

Singapore, however, has questionable human rights policies, does not have welfare laws and minimum wage for foreigners, which means hundreds of thousands of cheap labour temporarily immigrated can be imported from Bangladesh, Indonesia, and the Philippines. Those people live in camps and work for less than twenty dollars a day. Australia on the other hand has archaic and parochial labour laws which would require a minimum payment of $A200 dollars a day for foreign labour.

Business ethics and mankind motivation needs to shift a little more to the left where personnel greed and a large bottom-line profit at any cost may take a back seat to other community values like a cleaner environment, a safer and better workplace, a friendly happy workplace, higher employment, use of renewable resources, long term financial and standard of living gain for everyone in the company. Do business executives really need ten million per year? I have worked without pay for two years. Much of our economic system is linked too much to greater sales and bigger markets that require increased populations and growth.

Is business growth really all that important as it is in the unregulated capitalist system. However, when there are few resources left and growth cannot continue as before a new concept economic system needs to emerge. There will always be a demand for new or better products. As long as the company is making a profit and its employees and owners have good lifestyles, do we need anything else? Do we really need to raise our business executives 5 million per year because they closed down a shoe manufacturing plant in Thailand where the labour cost was $3 per hour in favour of one in China where the labour cost was $1 per/hour. Especially when they sell the shoes for $100 a pair that only cost them $3 a pair to make. It is this business practice glut and greed that has spread across America. In a world where local populations are diminishing the market should be aimed at, in addition to the new emerging third world country markets, creating better more economical products with better technology through government subsidised research and development.

Smaller populations are not a bad thing for the economy. For example, in Japan with all the tiny little homes, two homes can be turned into one larger nice home thus creating work and a better standard of living. This is happening in Houston, USA where older smaller homes are being pulled down, the lots amalgamated and one large home being built.

6.2 Reduction in Resource Exploitation

We will have to change and reduce drastically our consumption of new resources and commodities. Our energy resource oil is running out. The unstable price subject to the whim of OPEC has already wreaked havoc. What needs to be done now with minimum cost and change in lifestyle is the following.

Governments need to put a large import tax on oil and use the funds in a way that creates long term employment for all citizens and reduces dependency on foreign oil. The best solution is to first get the big gas guzzlers off the street. There is an abundance of coal and natural gas in the world. The standard fuel injected engine 4, 6, or 8 cylinder can easily be converted to run on natural gas (methane.) The price of methane is less than 22 cents per equivalent litre of petrol (gasoline) or 80 cents per US gallon. Methane is the cleanest burning fuel and can be produced by the anaerobic decomposition of any organic matter. Many garbage dumps make and burn methane and generate electricity. You can even make it from human and food scrap waste at home.

There are several innovative alternate fuel CNG Companies springing up in the USA, although illegal. Brazil, Argentina, India and Europe have 8 million cars on the road burning these cheap clean fuels. In the USA there are almost no cars yet using these fuels because it is illegal. The EPA requires extensive emission testing of any system for all car models and in addition, $100,000 has to be lodged with them to do their own tests. With CNG being the cheapest burning fuel, the most abundant in the world and can easily be fitted to any car, you must clearly recognise the inadequacy of the US Government to represent their countries interest. This law is set up to only protect the rich and the big car companies.

We all know how the big car companies have failed recently to provide the American citizens what they need and want. It's time for the democrats to push for change and new laws that allows this great CNG system to move forward. Two companies OES CNG, one of my companies, and Fuel Maker actually make home fill appliances and you can fill your car with CNG at home. Fuel Maker was recently purchased for peanuts by an oil rich tycoon named Boone Pickens whom I have not yet had the pleasure of meeting. I understand Mr. Pickens wants to run Texas electrical power on wind turbines and automobiles on natural gas. Pretty amazing and commendable for someone especially a Texan whom has gotten rich on oil.

The automobile especially after about 300 years when our natural gas supplies become tight, needs to be slowly phased out it is just too much of an extravagant machine. Mass transit with city trams and trains between cities in the USA needs to be reinstated to its 1930 glory period. In Australia I rode trams to work to the city of Melbourne for seven years. It was much cheaper than driving the family car. We saved the car for the weekends and the whole family rode. Electric trams are also like hybrid cars in that when you break the energy gas into the electric grid or battery in case of hybrids, instead of into wasted heat energy into the breaks. Electricity can also be generated from coal, gas and nuclear energy. Trains and trams carrying large groups of people are much more efficient; even if powered by oil-fired or steam-powered plants, they could cut US demand for foreign oil by 70%.

The Detroit automobile companies, steel manufactures, power companies, local governments, survey, civil engineering, and electrical supply and construction companies could all get in involved and all would have more than twenty years of growth and there would be little or no unemployment associated with this mass exodus away from foreign dependence on oil.

America largely consists of hundreds of small cities and towns with less than 500,000 people. To make mass transit universally applicable many people will have to move to the big cities or link up small towns with a train system with residential areas near train stations. The way it used to be and still is in Australia and most European cities. Freight will be much more economical and trucks transport limited to local areas. Getting large semitrailer trucks off the roads will reduce drastically the very expensive cost of road maintenance. Rail maintenance is much cheaper and so all the aging infrastructure in the USA would only have to be targeted where it supports the new trains and trams.

The recycling of consumer trash including glass, plastic, paper and metal cans will have and does have in many countries a tremendous effect in reducing the demand on our natural resources. In Australia everyone has 3 bins that are rolled out on the street, one is for compostable garbage, one is used for paper, plastic cartons, glass and cans and the other bin is used for garden cuttings. The garbage goes to tips and is used to make methane gas to power generators to make electricity. The bin with the glass, cans and paper goes to a recycling station where it is all sorted and sold to paper plastic, glass and metal manufacturers. The garden cuttings get ground or mulched and sold for lawn or garden top soil supplements.

6.3 Increase Renewable Energy Utilization

Renewable energy is a great concept but may take a hundred years or so and will require a large reduction in human populations, a concentration of people in large cities where hydroelectric, wind or renewable forestry products can be exploited for energy. There needs to be a lot more government sponsored research grants and tax incentives. For example, power generation plants can possibly be made in the oceans using millions of old cars and scrap metals as anodes. The benefit is three-fold, we get cheap energy, get rid of all the junk as the cars will dissolve and they don't have to waste energy melting them for recycling. Gold, platinum and other precious metals can actually plate out of sea water on the cathodes and a new cheap mining methodology is also born.

As time goes by hydrocarbon prices will increase including coal and then and only then will renewable energy sources like solar, hydroelectric, wave energy, geothermal and bio-fuels start to take over. But this will take 100–150 years and we first will need to ween ourselves from the automobile, reduce our populations, concentrate our populations in renewable energy areas and have much more energy efficient transport, home heating and lighting.

Bio fuels cannot be made from our food, like corn, wheat, or soybean or our heating, lighting, and travel will compete with our food. Food will win out every time. The best bio fuel is methane not methanol. It is easier to manufacture methane and it even can be made from rotting garbage, manure, almost anything. It can easily be produced at home by having all garbage disposal wastes and human wastes from the toilet going to an underground insulated holding tank kept at a constant temperature and without oxygen. Water levels can be pumped out into drainage fields automatically. The gas can be compressed and stored and used to burn as heating, cooking or to power your car. You can even throw in the lawn cuttings. This will allow you to drive your small economy 4-cylinder car about 20 kilometres (12 miles) per day for shopping when you are too lazy to take the train.

6.4 Cultural Globalisation

International awareness through television movies, documentaries and internet will eliminate a lot of energy intensive travel. Only the very rich or business executives will travel by air. Essential air travel will soon return to the days like it did in the 1950s.

The world will need to be ideally one day global "one world" not individual countries with trade tariffs. A global standard tariff of 10% if the product is made in the respective country should apply if it is not made in a country why have a tariff on it.

The United Nations needs to set standards on the quality of world leaders that ensure they have a global awareness and agree to adopt a democratic state. All leaders should have to be trained in the main cultures and especially religions of the world including Christianity, Buddhism, Muslim and Hinduism. All religions and schools need to acknowledge the peaceful coexistence of every other religion. There may even have to be a global integration like there was in the 1960s when black and whites in the USA were forced to live together. White students were bussed to predominantly black schools and black students to predominantly white schools. It worked and both races are much more tolerant of each other now days. We even have a half black half white president. America is growing up.

Singapore has a rich mixture of all religions makes it a law that only a limited number of people of the same faith live in the same neighbourhood. Personally, I would like to see all existing religions abandoned except Buddhism and a new one established. Buddhism is the only really peaceful, compromising and understanding religion of them all. The others embed too much past hatred and tyranny and you will never get Muslims and Hebrews, Muslims and Hindus or Muslims and Christians to forget the past and get on with their lives. I remember growing up in Miami, Florida, the Jews owned South Beach and to them, I, the young blond, blue-eyed, looked like a German, and I was not allowed in their stores and restaurants. Blacks occupied most of Liberty city, Coconut Grove and Fort Lauderdale. A white man didn't dare risk going there at night and have his car break down for fear of being robbed or murdered. I remember in 1960 a white redneck lived across the street from my family in Miami Springs and bragged how he used to go out and shoot niggers in Alabama.

I was spat on and given the rude finger by Muslims in 1995 in Saudi Arabia and nearly machine-gunned in Jisan, when I accidently walked into a prohibited zone near the port.

American white Christian radical rednecks in the Bible belt USA like the Ku Klux Klan could best be forced to relocate to Saudi Arabia and radical fundamental Muslims should best be settled and mosques built in Alabama, Texas, and North Louisiana, the biggest red neck capitals of the world. I remember in 1985 being invited to a luncheon at an exclusive golf course estate in Birmingham, Alabama by a group of small-time business executives concerning possible investment in my innovative oil and gas service company. The waiters and waitresses were black and only called boy or girl respectively. The south had truly moved ahead since the '60s as no one called them niggers in public anymore. A new world order possibly executed by the United Nations needs to not only act as an International Police Force counter acting acts against humanity but should help promote global awareness and cultural integration.

Most Muslims have adopted Western Culture are not fundamentalist and most Americans and Europeans are not white racist Christian fundamentalists. We need to root out the uneducated over religious trouble makers on both sides. If we can wipe out small pox and the plague, we can wipe our racism, terrorism, and religious fundamentalists.

7.0 The Electric Car Joke in 2021

Our Current presidency the Biden Administration is about to bankrupt the USA by excessive endorsement of alternative energy over fossil fuels and also promoting the Electric Car based on falsities like "Global Warming is caused from Carbon Dioxide," which you now know is false. They want to try and get our use of fossil fuels down 50% within twenty years and are endorsing the electrical car spreading lies that it runs on clean energy.

To me an Energy Engineer this is so false it is hard to believe a president of the United States of America would say something so stupid. I mean don't get me wrong I am a registered democrat and believe in getting rid of all hand guns and automatic weapons and other things pushed by the democrats. I also believe we have the best schools, roads, electrical power, and water and sewerage systems in the world and they are, guess what, socialised.

In America our electrical power comes from 40% very dirty coal, 27% natural gas, 27% nuclear and 6% from renewable sources like wind, Hydroelectric and solar. So, you can see 67% is from fossil fuels and coal is the dirtiest of them all. Coal has 50% more carbon than natural gas and 25% more carbon than gasoline. Because the other 33% of power sources has no carbon if you do the numbers however you find electrical power emits 85% of the carbon emission of natural gas and 68% of the carbon emission as for gasoline engines. Whilst there is less carbon emitted it is not that much and of course it is coal that will probably be used to make up additional power if the electrical car becomes wide spread and so these numbers will change.

If you check your power bill and convert to dollars to calorific value you will see electricity costs $3.50 to $4.00 per gallon compared to gasoline at $ 2.50 to $2.75 per gallon. Remember electricity has to be pushed through maybe 500 to 1000 miles of wires losing energy along the way.

Our existing electrical infrastructure will not take the charging of a significant number of electric vehicles without a complete rebuilding of our power stations, high voltage transmission lines and lines and transformers in your neighbourhood. My estimate is this will cost more than 10 trillion dollars or about $100,000 per household.

As far as alternate energy goes, it's a great idea but our current lifestyles won't allow it for most Americans. If you want to run a 1200-watt hair dryer twenty-four hours per day, on average in the USA you need 180 one square meter solar panels that are each putting out 20 watts and also storing in batteries for night time. The cost of the panels probably $18,000, cost of installation about $28,000 if you have a huge property and not in a hurricane area like Florida. If you have to buy suburb real estate allow another $200,000. The size area you need is about 360 square meters or about 3,873 square feet which is excessive for 90% of the people that live in American suburbs they just do not have that much area available. This is to run a hair dryer only and if you have air conditioning or electric cooking you will need more panels and more area. Solar Energy is only for people with a lot of money and live in the country and in areas with a lot of sun like out west or in Florida during clear skies.

I recently drove across Midwest America and was shocked at all the ugly tall windmill type wind generators. You can get small ones putting out 800 watts of power when the wind was blowing strong. You would need two of these to run your hair dryer or say four, assuming you could store the energy and you had suitable wind 50% of the time. Their cost is about $20,000 installed for all four of them. You would need to live in the country with a huge wind most of the time and as they are noisy and not only would your neighbours complain but it is probably illegal to install it just too dangerous. The fact in the winter of 2020, thousands of these windmills froze up in Texas and a million people had no power for a month should make you rethink windmill generators.

Hydroelectric is a good idea where you turn generators through water turbines. Look at the Tennessee River Valley Authority where they do just that. It's clean and somewhat economical but is only available for a limited number of people living in mountain regions like the Appalachian, Rockies, or Sierra Nevada Mountain ranges as you need a lot of water falling through a large height. At best 5% of the population could benefit from this power.

In summary, you have found out for the electric car:

1) Electricity is not clean most comes from the dirtiest fossil fuel coal.
2) About of a third of electricity comes from nuclear power and it may be dangerous I don't know it used to be.
3) Electricity has to be pumped up to 1000 miles through wires and loses efficiency
4) Electricity cost 35% more than gasoline in 2019.
5) To be able to power every car on electricity would cost three times the normal electricity costs as cars use more power. And this could be about $8,000 per year. In addition, the cost per household about

$110,000 to upgrade all the power stations, high-voltage lines, and lines to your house.

6) Except out west or in country Florida solar power will never power a hair dryer or air conditioner or electric stove without spending hundreds of thousands of dollars on solar panels and real estate.

7) Wind Turbines or Hydroelectric can only put out small power in regions where it exists and maintenance costs and dependability are key issues.